Summary *ix*

1 **Bandwidth Supply and Demand in Today's Army** *1*
Background on the Army's Digitization Initiative *4*
Bandwidth Supply at Army Commands Today *5*
Bandwidth Demand at Army Commands Today *8*
Comparing Bandwidth Supply and Demand in 2003 *12*
Operation Iraqi Freedom *14*

2 **Bandwidth Supply and Demand in 2010** *19*
Bandwidth Supply at Army Commands in 2010 *19*
Bandwidth Demand at Army Commands in 2010 *20*
Comparing Bandwidth Supply and Demand in 2010 *23*

3 **Mitigating Mismatches Between Bandwidth Supply and Demand** *25*
Buy Better Radios in Greater Quantities *25*
Reallocate Currently Planned Spending *27*
Reduce Demand and Better Manage Persisting Mismatches *28*

Appendixes
The Army's Current Communications Initiatives *31*
Extrapolating Continuous-Flow Information Across Command Levels *37*
Compressing Data to Reduce Bandwidth Demand *43*

Glossary of Abbreviations *45*

Tables

S-1.	Effective Bandwidth Supply Versus Peak Demand in 2003, by Command Level	x
S-2.	Effective Bandwidth Supply Versus Peak Demand in 2010, by Command Level	xii
1.	Maximum Engineering and Effective Bandwidth for Typical Army Communications Equipment in 2003	8
2.	Number and Types of Radios Available at the Ops Nodes in Digitized Units During Peak Operations in 2003, by Command Level	10
3.	Maximum Effective Bandwidth Available to Army Operations Networks in 2003, by Command Level	11
4.	Peak Demand for Continuous-Flow Bandwidth in the Digitized Division's Operations Net in 2003	11
5.	Peak Demand for Continuous-Flow Bandwidth in 2003, by Command Level	12
6.	File Sizes of Episodic Throughputs in Army Ops Channels in 2003, by Type of Throughput and Command Level	13
7.	Equivalent Peak Continuous-Flow Bandwidth for Episodic Throughputs in 2003, by Throughput Type and Command Level	14
8.	Total Peak Demand for Effective Bandwidth in 2003, by Command Level	15
9.	Effective Bandwidth Supply Versus Peak Demand in 2003, by Command Level	16
10.	Effective Bandwidth Available to Army Operations Networks in 2010, by Command Level	20
11.	Number of Operating UAV Systems, by Command Level	21
12.	Peak Effective Bandwidth Demand in 2010 Under Two Assumptions About TUAVs, by Command Level	22
13.	Effective Bandwidth Supply Versus Peak Demand in 2010, by Command Level	23

A-1.	Planned Investment in the Joint Tactical Radio System, by Organization and Cluster	33
A-2.	Effective Bandwidth Supply Versus Peak Demand in 2010, with a JTRS Substitute	35
B-1.	Peak Bandwidth Demand at the Digitized Division, Brigade, and Battalion Levels in 2003, by Mission	38
B-2.	Peak Bandwidth Demand for the Operations Nets at the Digitized Division, Brigade, and Battalion Levels in 2003, by Source of Demand	39
B-3.	Effective Bandwidth Supply Versus Peak Demand in 2003 Using a Factor-of-Two Extrapolation, by Command Level	40
B-4.	Effective Bandwidth Supply Versus Peak Demand in 2010 Using a Factor-of-Four Extrapolation, by Command Level	41

Figures

S-1.	The Army's Bandwidth Bottleneck	xi
1.	Notional Throughput Capacity per Node	26
2.	Total Projected Investment in Ops Net Equipment, by Command Level	27
3.	Cost of the Capacity Provided per Ops Net, by Command Level	27
4.	Cost of the Capacity Utilized per Ops Net, by Command Level	28

Boxes

1.	Information Technology: Some Concepts and Definitions	2
2.	The Evolving Mix of Wired and Wireless Battlefield Communications	6
3.	Operational Versus Theoretical Bandwidth	9

Summary

Bandwidth is a term used in much of the telecommunications industry as a measure, usually expressed in bits per second, of the rate at which information moves from one electronic device to another. To the extent that people are aware of bandwidth issues in everyday life, they most often confront them in the form of a shortfall in bandwidth—awaiting retrieval of an (Internet) Web page over phone lines and modems that are too slow, for example, or being told on Mother's day to phone again later because no telephone lines are available.

Bandwidth is a central issue for any communications system. The U.S. Army's current battlefield communications system still largely reflects the days when the pressing need was for verbal messages that relied on battlefield telephones. The expectation that a phone system would suffice changed in 1991, when then Army Chief of Staff Gordon Sullivan began the service's digitization initiative. The initiative's goals were to significantly increase Army units' knowledge, in real time, of the disposition and combat capabilities of both friendly and enemy forces, thereby providing commanders with a common operational perspective, and to improve their ability to command and control their forces by speeding up the delivery of information. General Sullivan's vision was to exploit the ever-increasing speeds of the microchip—which, in practice, meant introducing onto the battlefield unprecedented quantities of computers, communications equipment, and software.

An explosive demand for bandwidth accompanied the initiative. Underlying it was an expectation of responsiveness—that information would be transmitted quickly. The (often unarticulated) model was the increasingly fast Internet service that the civil telecommunications industry was providing. For the most part, the demand for bandwidth in homes and offices was met in the mid-1980s and early 1990s by what was at the time redundant telephone capacity predominantly through copper wires, which had accumulated over almost 100 years. As that capacity was exhausted in the 1990s, the telecommunications industry increased the available bandwidth by laying thousands of miles of fiber-optic cable—a solution that, for the most part, is not available to Army troops on the move during battle.

Some personnel in the Army recognized that the bandwidth provided by the battlefield telephone system would be inadequate to support the goals of digitization. Nonetheless, the service did not fully realize the challenge that implementing digitization would pose to both its communications community and the wireless telecommunications industry in general. In particular, the Army did not know the size of its total requirement for bandwidth and had not considered the substantial difference in cost between using fiber-optic cable and using radios to satisfy that demand. (For an equal amount of bandwidth, cable is currently about 25 to 50 times cheaper.)

After a decade of effort, the Army's digitization initiative has fallen short of its goals;[1] the service's plans now focus

---

1. The Army planned as recently as 1999 to bring digitization to 47 brigades (32 active and 15 reserve) by 2009. As of now, however, it has abandoned its intention to digitize more than a two-division corps plus its support elements and is now focusing on digitization of the interim portions (the Stryker brigades) of the transformed Army's future Objective Force. The new goal is to digitize 13 units equivalent to brigades by 2009.

on transformation, although digitization still remains an objective, albeit a subsumed one. (Transformation aims for significantly lighter and more easily deployable units that nevertheless have increased lethality and survivability compared with units today.) Consequently, the Army continues to invest significant amounts of money in new communications bandwidth to support its revised goals for digitization as well as new programs associated with transformation. The latter category includes the planned widespread use of unmanned aerial vehicles (UAVs), whose operation requires large amounts of bandwidth.

In this study, the Congressional Budget Office (CBO) analyzes the current and future total demand for communications bandwidth to support operations officers at all levels of command within the Army. CBO then compares that demand with the total bandwidth supplied by communications systems in place today and those planned for the future. Although numerous studies of Army communications bandwidth have been conducted, none has attempted to assess total demand and supply at all levels of command.

CBO's analysis is intentionally conservative insofar as it attempts, when there is doubt, to assess both future communications capabilities as being greater than they might be and future bandwidth demand as being less than it might be. Therefore, in cases in which CBO projects that future demand appears to exceed supply, this study may underreport the degree of mismatch.

CBO's analysis has yielded the following conclusions regarding the bandwidth available to operations officers. First, at all levels of command within the Army, the current demand for bandwidth is larger than the supply—shortfalls of as much as an order of magnitude (or up to 10 times the amount of supply) can exist. Second, shortfalls in the supply of bandwidth will persist at some command levels through and after 2010, when the capabilities associated with the Army's transformation begin to be put into the field. Thus, after what is now planned as an investment of approximately $20 billion in new communications equipment, the Army will fall short of its goals at certain command levels by an order of magnitude. (Unless otherwise indicated, all dollars are fiscal year 2003 dollars.)

Bandwidth Supply and Demand in Today's Army

At various Army headquarters in the field—for example, at the corps, brigade, or division levels of command—there are multiple communications networks. This analysis focuses on networks that serve the operations officers, denoted the *ops desks* or *ops nets*. Other communications networks (often located in the same command centers) are not unimportant, but in many cases they have similar equipment and are used to transmit similar amounts of information. (Examples of other communications networks are the intelligence and fire-support nets, which serve, respectively, intelligence personnel and personnel in the headquarters who control attacks on targets.)

Summary Table 1.
Effective Bandwidth Supply Versus Peak Demand in 2003, by Command Level

Command Level[a]	Relative Supply Versus Peak Demand (S : D)[b]
Corps	1 : 1 to 4
Division	1 : 5 to 8
Brigade[c]	1 : 1.5 to 3↑
	1 : 20 to 30↓
Battalion	1 : 10 to 20
Company	1 : 2 to 6
Platoon	1 : 0.5 to 2
Squad/Vehicle	1 : 2 to 6

Source: Congressional Budget Office.

a. At the higher command levels, the table refers to the operations networks only. At lower levels, the distinctions between the various communications networks (for example, operations, intelligence, and fire-support) become less clear.

b. Based on an approximate logarithmic scale, the color coding is as follows: yellow indicates that supply is between about one-third and three times demand (a marginal supply/demand match); light orange signifies that demand is approximately three times supply, and orange, that demand is between three and 10 times supply. Red (used here for the lower brigade-level relationship and at the battalion level) means that demand exceeds supply by a factor of 10 or more.

c. The up-arrow (↑) indicates the throughput (bandwidth) rate for communications to equivalent or higher command levels. The down-arrow (↓) indicates the throughput rate to lower command levels.

Summary Figure 1.
The Army's Bandwidth Bottleneck

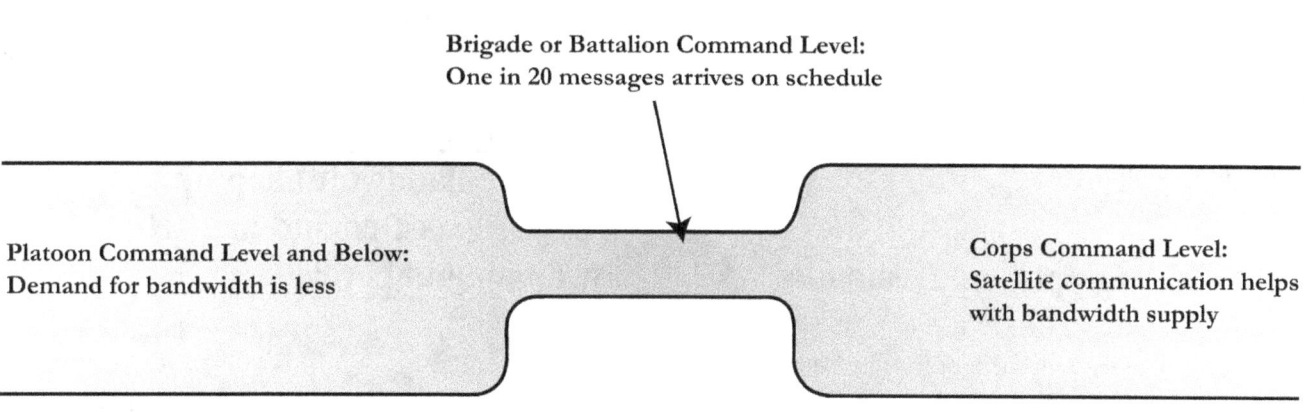

Source: Congressional Budget Office.

Therefore, CBO's analysis of bandwidth supply and demand for the ops nets may illuminate supply/demand mismatches occurring in the Army's other communications networks.

Expressing the supply of bandwidth at each command level as one, Summary Table 1 displays the current relationship between supply and demand. Ranges are used for peak demand because only rough estimates of it are available. (The Army's data on peak bandwidth demand are a limited collection of experimental observations, test results, and simulations that have not all been obtained in consistent, controlled environments. In view of those limitations, CBO has reviewed its use of the available data with communications experts throughout the Army.) CBO's analysis focuses on peak demand because, with rare exceptions, it corresponds to the most intense battlefield activity, when failure is least tolerable.

The color coding in the table indicates how well the estimates of bandwidth supply and demand match. At no level of command is there a significant excess of supply relative to demand. If supply and demand are within a factor of about three (based on an approximate logarithmic scale), yellow is used to express caution. Caution is warranted for two reasons: modest delays in transmission are likely, with unknown effects on operations; and there is little potential to accommodate additional growth in demand.

The colors light orange, orange, and red indicate worsening degrees of mismatch. Light orange and orange correspond to increasingly long delays in message transmission or lost messages. Red—used in Summary Table 1 for the mismatch at the brigade level indicated by the down-arrow and the mismatch at the battalion level—means that demand exceeds supply by factors of 10 or more, resulting in severe operational effects. Computer memories, file systems, and network file servers overfill and sometimes "crash"; the network, or parts of it, fail; and command and control becomes "analog" as soldiers abandon the digital system and return to voice-only communications. (See Summary Figure 1 for a simplified depiction of the mismatch between bandwidth supply and demand that currently exists at different levels of command.)

The reversion to analog command and control is an outcome reported both officially and unofficially in many of the Army's advanced warfighting experiments (AWEs) that were conducted between 1998 and 2001. The recent war in Iraq, by comparison, saw troops adopting a different approach. The Army provided them with a limited version of digitized communications (which nevertheless included the new capability of graphical displays of cumulative data), and units handled bandwidth shortfalls by using character-based transmissions (which are similar to ones from the terminal teletypes of an earlier era). Those messages were transmitted using e-mail and "chat room" formats (like those available over the Internet), which demand relatively little bandwidth.

Bandwidth Supply and Demand in 2010

Within the decade, both the supply of and demand for bandwidth on the battlefield are expected to increase substantially with the fielding of components of the next-generation Army. Key among them are the first units to be outfitted with the Future Combat System family of vehicles, the Shadow UAV, and a large amount of new communications equipment. The programs now generating that equipment are major elements of the Objective Force, the Army's current name for the lighter, more deployable force that it expects to have in place in 2010.

Advances in communications equipment that the Army plans to field include the Joint Tactical Radio System (JTRS), Warfighter Information Network-Tactical (WIN-T), and Multiband Integrated Satellite Terminal (MIST). The WIN-T and MIST are major components of the Army's plans to improve communications between the brigade, division, and corps command levels (known as the upper tactical Internet). The JTRS is the primary advance anticipated in communications between the brigade command level and smaller units (known as the lower tactical Internet). A number of features differentiate the two Internets; an important one is that the upper tactical Internet is augmented with satellite communications, whereas the lower is not.

In addition to the new communications equipment, the Army is considering a change in the internal architecture of its communications networks, which by 2010 would redistribute the information that those networks carry. A plan being discussed for the Objective Force would load all non-UAV information from the operations, intelligence, and fire-support networks onto a single "backbone net" and divert all UAV data to a separate high-capacity network.

By 2010, as *Summary Table 2* shows, the most severe mismatch between supply and demand will have moved

Summary Table 2.
Effective Bandwidth Supply Versus Peak Demand in 2010, by Command Level

Command Level[a]	Relative Supply Versus Peak Demand (S : D)[b]
Corps[c]	1 : 10 to 30
Division[c]	1 : 10 to 30
Brigade[c,d]	1 : 3 to 10↑
	1 : 5 to 15↓
Battalion	1 : 1.5 to 3
Company	1 to 4 : 1
Platoon	4 to 10 : 1
Squad/Vehicle	7 to 20 : 1

Source: Congressional Budget Office.

Note: The range in relative demand at the brigade and higher levels of command is associated with either the proposed "backbone" architecture or the case in which the current architecture is maintained and downlinks from unmanned aerial vehicles are heavily networked. (See Chapter 2 for details.)

a. At the higher command levels, the table refers to the operations networks only. At lower levels, the distinctions between the various communications networks (for example, operations, intelligence, and fire-support) become less clear.

b. Based on an approximate logarithmic scale, the color coding is as follows: green (used here at the platoon and squad/vehicle command levels) means that supply exceeds demand by approximately a factor of three or more. Yellow indicates that supply is between about one-third and three times demand (a marginal supply/demand match), and orange signifies that demand is approximately three to 10 times supply. Red (used here for the corps and division levels) means that demand exceeds supply by a factor of 10 or more.

c. If the architecture for information distribution located in the tactical operations centers at these command levels is altered, as the Army is considering doing, then operations officers will share information in a new "backbone net." In that case, the upper end of the projected range of bandwidth demand would apply.

d. The up-arrow (↑) indicates the throughput (bandwidth) rate for communications to equivalent or higher command levels. The down-arrow (↓) indicates the throughput rate to lower command levels.

from the brigade level to the corps and division levels (shown in red on the table). That shift occurs because the JTRS, if it works as planned, will significantly increase the supply of bandwidth relative to demand in the lower tactical Internet. Further, at the lowest command levels, the JTRS will provide some capacity to accommodate additional growth in demand beyond 2010.

At the higher command levels, however, the increase in the supply of bandwidth attributable to the WIN-T and the MIST will be swamped by substantial increases in demand, CBO projects. As a result, the new backbone nets, if implemented, will experience bottlenecks at the corps and division levels that are quantitatively similar to those existing at the battalion and brigade levels today.

Options to Mitigate Mismatches Between Bandwidth Supply and Demand

CBO examined several options to either improve the future match between bandwidth supply and demand or lessen the risk that the mismatch will be significant. Its analysis considered three general approaches: boosting the amount of bandwidth above the increases already envisioned in the Army's programs, reducing the demand for bandwidth, and better managing the mismatch between supply and demand.

CBO chose not to develop alternatives that would increase the supply of bandwidth after it analyzed two such approaches: funding efforts to develop new technology that would mature by 2010 or buying more of the currently planned systems. The technology approach is probably not feasible because the Army's planned new communications programs are already adopting all current and projected advances in technology. An approach that increases capacity by purchasing more equipment is also problematic. For example, some Army experts have suggested that projected bandwidth demand can be met by purchasing 20 times more sets of WIN-T equipment than the service is now planning to buy. However, current cost estimates for the WIN-T program already range between $4 billion and $9 billion, and additional spending of that magnitude may simply be an unlikely prospect.

The three alternatives described below suggest ways to eliminate lower-priority demand for bandwidth and better manage the demand that remains. The first two alternatives would stop the transmission of information that might be of lesser priority yet would be expected to contribute significantly to the future mismatch between bandwidth supply and demand at some command levels. The third would mandate the adoption of software tools, some of which are starting to become commercially available, that could allow better management of the demand for bandwidth when it exceeded supply.

- Eliminate video teleconferencing, and in its place substitute teleconferencing plus a whiteboard capability.

- Do not network UAV "downlinks." Maintain the autonomous sensor-to-shooter links (video downlinks from the UAVs), but do not transmit the video signals that they produce throughout the command structure, as the Army has planned. (Under this option, UAV video feeds would be sent directly to a single fire-support or intelligence center but would not be transmitted farther.)

- Mandate that automated bandwidth management tools, which dynamically constrain the use of bandwidth and are now coming onto the commercial market, be built into the software that is expected to underlie both the upper and lower tactical Internets in 2010. The earliest such tools focus on network software; by implementing them and later extending them to applications software, the Army could manage future bandwidth shortfalls as they occurred and mitigate their effects.

CHAPTER 1

Bandwidth Supply and Demand in Today's Army

The Office of the Joint Chiefs of Staff has asserted, in its 2001 publication *Joint Vision 2020*, that "[i]nformation, information processing, and communications networks are at the core of every military activity." In response to that guidance, the Army is seeking to create an information technology (IT) environment in which computers are routine battlefield tools, information (measured in bits per second) passes continuously over wired and wireless channels, and the complex communications infrastructure necessary to keep all of that gear operating is in place. This Congressional Budget Office (CBO) study focuses primarily on how the Army transmits streams of information, known as bandwidth, and how much information it wants to transmit, both now and in the future. (*Box 1* defines and discusses several IT terms and concepts.) In particular, the study addresses mismatches between the amount of bandwidth demanded at various command levels and the amount of bandwidth supplied.

The Army faces a number of problems in implementing its IT strategy on the battlefield. The service needs much more bandwidth than it has available today to support both its current systems and those planned for the future, but the cost of additional bandwidth is high. Moreover, as this study concludes, the Army's planned investments in communications capability will not fulfill its projected future requirements. Those conclusions have been echoed in numerous discussions that CBO has conducted with members of the Army's IT community as well as in eight recent studies on various aspects of the service's demand for bandwidth.[1] None of those studies, however, has attempted to systematically assess the Army's total band-

---

1. Monica Farah Stapleton and Yosry Barsoum, "C4ISR Systems Engineering Analyses and Modeling & Simulation" (briefing prepared for the Army Communications and Electronics Research, Development, and Engineering Center by AMSAA [Army Materiel Systems Analysis Activity] and the Mitre Corporation, September 9, 2002); J.L. Burbank and others, "Concepts for the Employment of Satellite Communications in the Army Objective Force" (draft, Johns Hopkins University Applied Physics Laboratory, August 2002); RAND Arroyo Center, "Future Army Bandwidth Needs—Interim Assessment" (briefing prepared for the G6/Army Chief Information Officer, July 10, 2002); Yosry Barsoum, "Bandwidth Analysis (ACUS Only) of Division Main, Maneuver Brigade TOC, and Tank Battalion" (briefing materials prepared for the Army's Communications and Electronics Command by the Mitre Corporation, February 29, 2000); Steve Chizmar and others, "Digitization at Brigade and Below (DB2) Study" (briefing prepared for the Army Materiel Systems Analysis Activity, Aberdeen Proving Ground, Maryland, December 1, 1999); Army Medical Department Center and School, "Medical Force Digitization Overview" (briefing prepared for the Deputy Chief of Staff for Command, Control, Communications, and Computers, January 1999). The following were provided to CBO by the staff group of the Army Chief of Staff: Army Signal Center and Fort Gordon, Directorate of Combat Developments, Modeling and Simulation Branch, Architecture Division, *First Digitized Forces System Architecture (1DFSA): Version 2.02, Simulation Analysis/Study*, White Paper (main text and attachment titled "Satellite Communications Capacity Study," June 30, 1999; released to CBO on May 14, 2002); and Lt. Gen. P. Cuviello, "Projected Bandwidth Usage and Capacity" (briefing prepared for the Army Chief of Staff by the G6/Army Chief Information Officer, August 2002; released to CBO on January 14, 2003).

Box 1.
Information Technology: Some Concepts and Definitions

To allow a better understanding of the analysis and conclusions presented in this Congressional Budget Office (CBO) study, some common terms and concepts from this technical field are discussed below.

Measuring Information Flow

The rate of flow of information is usually expressed in *bits per second*, a bit being the smallest representation of information in a binary code of zero or one. A shorter phrase for bits per second is *bandwidth*, as defined in the applied communications industry.[1] This report uses bandwidth synonymously with the term *throughput*.

The more bits per second that an information flow contains, the higher the bandwidth. Convenient units for bandwidth today are typically kilobits per second (thousands of bits per second, or Kbps), megabits per second (millions of bits per second, or Mbps), or gigabits per second (Gbps), which is a thousand times more. Most communications equipment today operates at rates ranging from a few kilobits per second to one or more gigabits per second.

Voice Transmission Rates. For the spoken word, actual information content flows at hundreds to a few thousands of bits per second. For example, the phrase "There's a fire in the room out back" is usually spoken in about a second. That is, in a second, 35 characters, each represented by eight bits of information (in the ASCII code, for instance), or some 280 total bits, are communicated. Very rapid speakers, such as auctioneers, might convey 3,000 bits of information per second.

When voices are transmitted over a phone system today, much more than just the core information is conveyed; the transmission also includes tone, volume, timbre, and other nuances that distinguish one voice from another. After a century of engineering, the number of those additional *context bits* per bit of core information is about 10 times more. Consequently, commercial phone systems do not operate at 3 Kbps but are instead usually rated at a minimum of 32 Kbps per phone line. The Army standard is somewhat lower—16 Kbps per phone line—but is essentially as satisfactory for most users.

Data Transmission Rates. Although people who talk on the phone are content with the level of communications provided by 32 Kbps of bandwidth, they demand much higher throughputs when using their home computers on the Internet. Today, many people who access the Internet over a phone line use a 56 Kbps modem. (A modem is a device that converts signals produced by, say, a computer to a form compatible with, for example, a telephone.) Fifteen years ago, a typical speed for a modem was 1,200 bits per second. As modem speeds began doubling, and doubling again, 56 Kbps was the first modem speed that took full advantage of the rated throughputs of standard phone lines and telephone interfaces—as well as nearly accommodating the first level of improvement above 56 Kbps, which is usually 64 Kbps.[2] (Faster speeds require alternative technology,

---

1. That definition of bandwidth is not the one used in scientific fields or in much of the engineering community. There, bandwidth is defined as the difference between two frequencies (radio frequencies, for instance) rather than bits per second. However, the two definitions can be functionally related and are nearly proportional at low noise levels. Thus, communications specialists often shift back and forth, without confusion, between the two definitions in different technical contexts.

2. The actual throughput of a typical copper point-to-point commercial phone line varies and depends not only on commercial switching rates but on the quality of the telephone as well. Although most phone lines are conservatively rated at 32 Kbps, they can operate easily at 64 Kbps when they have been equipped with higher-quality interfaces.

Box 1.
Continued

such as a digital subscriber line, or DSL, or a cable system that bypasses the phone system's copper wires.)

Video Transmission Rates. As was the case with the above analysis of phone lines, it is relatively simple to crudely estimate the size of the bandwidth required for transmitting video images. The typical computer screen today is about 1,000 pixels across by 1,000 pixels high. (Pixels are the small, individually modifiable elements of a video screen.) Basic color representation requires a minimum of eight bits of information per pixel. The cinematic illusion of movement requires about 32 frames per second.[3]

In addition, just as the phone system requires an extra order of magnitude of bandwidth for the context bits, so, too, computer systems require context bits in about the same 10-to-1 ratio. (The context bits in video transmissions, however, handle such features as noise, packet switching, error correction, and retransmission.) Therefore, if no adjustments were made, the bandwidth required for a single continuous video transmission (sometimes called streaming video) would be about 2,560 Mbps [2,560 bps = (1,000 x 1,000 pixels/frame) x (8 bits/pixel) x (32 frames/second) x (10 context bits/bit)].

Fortunately, modern engineering practices reduce that information load by about a factor of 100 by applying data compression and so-called delta techniques.[4]

Moreover, trimming the number of pixels in the frame slightly and the frames per second somewhat can reduce the load further, by a factor of 10. Therefore, with a penalty of only a slight loss in resolution and an increase in flickering, 2,560 Mbps can be reduced to a bandwidth load of just a few megabits per second, which is typical for a single point-to-point video transmission.

Software Layers

In a computer network, the software necessary for its operation is sometimes differentiated into layers, and measurements of information bandwidth differ from layer to layer. An e-mail system is an example of *applications-layer* software. If an output—that is, an e-mail—is to be sent to another user across a large network, the outputs of the e-mail software must be augmented by information from local operating systems, network software, and numerous other software layers. If at some stage in the transmission a radio is employed, more software (the *data link* and *physical layers*) is needed at the lowest network layers to transmit bits over an antenna, monitor whether the bits are transmitted without interference, and initiate retransmissions if errors occur. A higher level of bandwidth is required at the data link and physical layers than the level needed to support a throughput at the applications layer.

This Congressional Budget Office study distinguishes between the two levels. Bandwidth associated with the data link and physical layers (the throughput that is often cited by hardware manufacturers) is called *engineering bandwidth*; bandwidth at the applications layer is referred to as *effective*, or *operational, bandwidth*. Specifically, CBO's analysis compares the demand for and supply of effective (or operational) bandwidth for the Army's point-to-point communications, because that level of throughput is what personnel observe in the field.

---

3. Some observers notice "jumpiness," or monitor flicker, if the speed of motion is much less than that. Commercial standards today are usually better, or about 60 frames per second.

4. Delta techniques are a set of algorithms for determining those portions of the new frame that are different from the last frame. Once determined, only those different portions are transmitted.

width requirements by communications network and command level.

In this study, CBO analyzes the current and future total demand for communications bandwidth to support operations officers at all tactical levels of command (from the corps down to the individual squad or vehicle) within the Army. The analysis then compares that demand with the total bandwidth supplied by the service's current and planned communications systems. This chapter considers bandwidth supply and demand today; Chapter 2 estimates supply and demand in 2010. Chapter 3 discusses several options for addressing the bandwidth supply/demand gap that exists today and is expected to persist—although at a different point in the command structure—in the future.

Background on the Army's Digitization Initiative

The Army's battlefield communications system was originally designed to handle voice and character-based (old-fashioned terminal teletype) messages. In general, those types of communications, whose upper limit on rate of flow is several kilobits per second, require relatively little bandwidth. To send 100 poor- to average-quality voice messages simultaneously, therefore, requires bandwidth in the range of several hundred kilobits per second. Rates for character-based information are similar. The difference between several and several hundred kilobits per second generally defines the range in bandwidth, or throughput, of the tactical radios that the Army employs today on the battlefield. (Certain augmentations to that equipment were fielded during the recent Iraq war and constitute exceptions to that statement. They are discussed at the end of the chapter.) Using mobile retransmission towers (relays) around the battlefield and the current generation of satellite communications, the Army can meet its current needs for bandwidth for voice and character-based communications.

To transmit computer data and video signals, however, requires much more throughput. Such transmissions on today's battlefields and those of the future include video "downlinks" from unmanned aerial vehicles (UAVs), computer displays of data depicting battlefield situations, video teleconferencing, and telemedicine, in which doctors examine wounds and advise on treatment remotely. Much of the bandwidth demand reflected in those examples is associated with streaming (continuous) video imaging, which requires a minimum of several megabits per second of throughput to be considered continuous by most viewers.

If 100 video and data transmissions are to be supported simultaneously—a typical scenario in a larger headquarters such as for a division or corps, the demand for bandwidth grows from several megabits per second to several hundred. That throughput is three orders of magnitude (10 x 10 x 10, or 1,000, times) greater than the Army's current communications system was originally designed to handle. It is more than 10 times larger than the demand that the system is projected to be able to handle in the future, following the improvements that the Army is currently planning.

The substantial increase in bandwidth demand on today's battlefield is due to the Army's digitization initiative. Begun in 1992 by then Army Chief of Staff General Gordon Sullivan, the initiative was designed to field advanced information technologies (computers, software, databases, and communications equipment) to rapidly provide "situation awareness" and support for decisionmaking. The phrase *situation awareness* denotes knowing the locations of both U.S. forces (in military terms, referred to as Blue situation awareness) and enemy troops (known as Red situation awareness). Within a short time, Army leaders had also identified the subsidiary goals of fielding a first digitized division by 2000 and a first digitized corps by 2004, with the rest of the Army to follow.

The digitization of the Army has proceeded more slowly than was originally anticipated. As recently as 1999, the Army was planning to have 47 digitized brigades (32 active and 15 reserve) in the field by 2009. However, when planning began in 1999 for the new Objective Force of the future (following the installment of General Eric Shinseki as Army Chief of Staff), the idea of carrying digitization beyond the first digitized corps was abandoned and supplanted by new goals for the "transformed" army, which incorporated the objectives of digitization as a subset. (The Army's transformation involves making forces deployable more quickly while maintaining or improving their lethality and survivability.)

Under the umbrella of the Objective Force, the number of brigades to be digitized by 2009 has been reduced to 13—seven maneuver brigades in the first digitized corps and six new Stryker brigades. (Stryker brigades are interim units whose organization and capabilities place them between traditional brigades and Objective Force units.) By the end of 2003, the Army plans to have five digitized brigades. Four will be components of the 4th Infantry Division and 1st Cavalry Division in the first digitized corps, and the fifth will be the first Stryker brigade.

Bandwidth Supply at Army Commands Today

To understand how the Army intends to manage the communications networks and equipment that it now has —and the bandwidth they provide—one must consider how the hardware being used and the information coursing through it are handled within the Army's tactical command structure. The model of that structure used by CBO in its analysis is the structure of the first digitized division (the 4th Infantry Division) and the first digitized corps (the Third Corps). By adopting the Army's recent digitization framework, or architecture, as its starting point, CBO's analysis follows the Army's current IT strategy fairly closely, noting anticipated future changes in architecture as appropriate. In addition, while battlefield situations involve both wired and wireless (sometimes referred to as radio frequency) communications, this study focuses more on wireless messaging, because of its dominance in mobile battlefield operations and its relative cost. (*Box 2* gives some sense of the interplay between the two types of messaging on the battlefield.)

In the digitized Army, radios are organized into communications networks. As a shorthand, the networks are often referred to, synonymously, as *nets* or *nodes*. Army nets can be differentiated by the combat mission (operations, intelligence, logistics, fire support, and so on) that they predominantly serve. They are also characterized by the types of radios that produce their transmissions and by the other nets with which they directly communicate.

For instance, the operations officer in a battalion will manage communications from the operations net, usually designated the "ops" net. Today, the radios that this officer typically uses include an NTDR (Near-Term Data Radio), an EPLRS (Enhanced Position Location Reporting System), and one or more SINCGARSs (Single-Channel Ground and Airborne Radio Systems). At the battalion level, the operations officer (or a subordinate) directs large volumes of information to operations officers in the companies below and in the brigade above; takes orders and assessments from the brigade; and receives and transmits situation assessments from other operations officers in adjacent battalions.

Lesser volumes of information also pass to the ops net from the intelligence, fire-support, and other networks.[2] When such nets are colocated in the same command center, communications between them pass over a wired network, usually a local area network (LAN).[3] When nets are not colocated, which is common for communications between different levels of command or between forward and main headquarters, information from the intelligence, fire-support, and other nets that must be shared places additional demands on the ops net's wireless bandwidth. (*Table 1 on page 8* presents the equipment—mostly radios —that is now found in the operations nets of digitized Army units. The new and improved radios and other equipment that the Army plans to develop and field by the end of the decade are discussed in Chapter 2.)

The effective communications bandwidth provided by the Army's voice and data radios is typically reduced below the maximum (engineering) rates by a factor of about 10, owing to the bandwidth needed for context bits and channelization (*see Box 3 on page 9*). Because the mobile subscriber exchange, or MSE (the equivalent of a private telephone system), has a much higher degree of channelization than any of the other radios, its throughput is reduced by roughly a factor of 100 for point-to-point transmissions.

The operations officer at a given level of command has one or more radios, or "pipes," available for communica-

---

2. Those networks carry information internal to their mission areas and analogous in size to the large volumes passing through ops channels.

3. A LAN is one example of a network that is "wired," or connected, by very high bandwidth fiber-optic cable.

Box 2.
The Evolving Mix of Wired and Wireless Battlefield Communications

Traditionally, battlefield communications systems have used radios (wireless communications), and very high bandwidth messaging made possible through the use of fiber-optic cable (wired communications) has been nonexistent. Wired communications using copper wires have long existed only in or near tactical operations centers (TOCs), with the principal purpose of tying together its local phone system. With such a system, the Army fought Desert Storm, and it faced no significant restrictions on its operations because of shortfalls in bandwidth.

In that conflict, the pace of the throughputs that radios provided for the command-and-control system (including what were then considered high-capacity satellite communications) was satisfactory because nearly all messages represented analog voice traffic for which the equipment and procedures were designed. U.S. forces were able to communicate much more freely than their adversaries could, and U.S. commanders' decision-making cycle was much faster as well. By the time of the war in Afghanistan, special forces personnel were using military variants of (wireless) cell phones to transmit voice and simple numerical data, such as global positioning system coordinates, to Air Force and Navy bombers.

In some battlefield situations in which the Army requires very high levels of bandwidth, it is available. Specifically, in combat environments where distances between operations centers are short, the nodes (in this instance, a group of colocated nets) move infrequently or not at all, and the risks of direct enemy attack are negligible, large demands for bandwidth are easily handled. In those circumstances, Army doctrine calls for running high-bandwidth land lines in the form of bundles of copper cable or using pulse-coded modulated (PCM) lasers, or both, to communicate between nodes. Also, within the TOCs of digitized units, fiber-optic cable is routinely used to connect elements of the internal local area networks, or LANs.[1]

Both PCM lasers and LANs support megabit-per-second throughput. But they are restricted in their geographic extent and limited to fixed nodes. The transmission range for a LAN is usually several hundred meters.[2] PCM lasers are limited to about 20 kilometers because of their susceptibility to beam drift caused by wind and other atmospheric conditions. LANs and land lines typically avoid that problem, but they can take several hours to deploy.

---

1. That kind of connection has been demonstrated within TOCs associated with the 4th Infantry Division. Although in principle, a variety of LAN architectures are possible, the Army's intra-TOC LANs are organized as 100 Mbps Ethernets. (An Ethernet is a kind of LAN that is now recognized as an industry standard.)

2. LAN geography is always limited, but the details depend on the speed of light in the fiber, the minimum synchronized operating frequencies of the nodes and other equipment on the network, the network's topology, and other physical characteristics.

---

tions (*see Table 2 on page 10*). The approximate capacity of what is termed a trunk line at a particular command level (essentially, the operations network) can be obtained by adding up the capacities provided by the available individual pipes.[4]

Combining the information in Tables 1 and 2, CBO estimated the maximum total effective throughput that the Army's ops nets can provide. To do that, CBO made several assumptions (which represent ideal circumstances): for a given point-to-point communication, the throughput load can be fitted to perfectly match the maximum oper-

---

4. If the demand for information through those pipes could be exactly matched to their size, then the trunk line's capacity would exactly equal the sum of the pipes' bandwidths. In practice, however, that does not always occur; therefore, the sum represents an upper

bound on the trunk line's actual bandwidth. The trunk line for digital communications does not include a number of older, generally analog radios that carry relatively little message traffic.

Box 2.
Continued

In addition to those communications techniques, the Army's operational headquarters in Korea, Saudi Arabia, and Europe were linked within the past decade to the Pentagon and other U.S.-based headquarters using fiber-optic-based international commercial networks. Often called wide-area networks (WANs), they have become highly redundant in much of the world during the past 10 years. From a practical standpoint, however, WANs are restricted to fixed nodes.

Such approaches (WANs, LANs, PCM lasers, or bundled land lines) are generally of high bandwidth, but they are not practical for mobile vehicles and are therefore inappropriate for maneuvering forces that operate over wide areas. They are also less useful in cases in which a given message has a large intended audience or the precise locations of the recipients are unknown. In such instances, traditional (wireless) radios are ideal. They operate in broadcast mode, sending their signals out widely, and are useful for all levels of forces. Consequently, wireless, in spite of its lower intrinsic bandwidth relative to wired communications, and its much higher cost (25 to 50 times more) per unit of throughput, is the most practical communications medium for troops.

To supply its deployable forces, the Army has fielded hundreds of thousands of radios. The number of nodes that must in principle be linked by wireless technology to support an army on the move is about 100,000. To provide largely voice communications, the Army's inventory of radios in the field numbers about 220,000, or enough to support two major theater wars simultaneously.[3]

Wireless technology can reach the megabits-per-second range, and the Army currently has several programs designed to bring that rate of bandwidth to the individual soldier. The problem is that such high-bandwidth radios, designed to operate both on their own and with existing radios, may cost about $127,000 apiece.[4] The minimum investment required to replace the entire fielded inventory of radios with new models is thus about $28 billion. In addition, because those replacement radios are still in development, that estimate of their cost is subject to a number of risks (discussed later), some of which could increase the investment substantially.

---

3. The figure 220,000 is also approximately the number of fielded radios capable of digital communications. In addition, the Army fields tens of thousands of older, mostly voice-only radios. However, the Congressional Budget Office has generally ignored those radios in its analysis because they are not part of the battlefield Internet and their contribution to overall communications bandwidth is small and decreasing.

4. That average cost was noted in the December 2002 Selected Acquisition Report for the Joint Tactical Radio System, Cluster 1. The average unit procurement cost including research and development spending exceeds $140,000.

---

ational capabilities of the available radios; input and output throughputs are equal;[5] and all intended recipients are within an average operational range of all the radios. Under such assumptions, the data throughput rates of individual radios can be added to estimate total throughput capability—the total bandwidth supply—provided by the radios available in each ops net (*see Table 3 on page 11*).

As the table shows, the digitized brigade has two measures of available bandwidth. That command level employs both the Army's beyond-line-of-sight (BLOS) satellite com-

---

5. For the most part, that assumption is valid for ops nets, which according to Army doctrine share information with both higher and lower command levels. For other nets, the assumption may not apply. Fire-support nets, for example, often require much higher bandwidth for inputs than for outputs because their incoming traffic is dominated by data from high-bandwidth sensors such as UAVs, whereas their outgoing traffic is made up of relatively low bandwidth messages telling troops to fire at specific targets.

Table 1.
Maximum Engineering and Effective Bandwidth for Typical Army Communications Equipment in 2003
(In kilobits per second)

Radio/Communications Equipment[a]	Typical Battlefield Command Levels	Point-to-Point Data Throughputs	
		Maximum Engineering	Average Effective[b]
SINCGARS (SIP)	Vehicle to Corps	16	1.7
EPLRS (VHSIC)	Company to Corps	128	13.3
NTDR	Company to Corps	288	30
Interface Standard[c]	Battalion to Army	16	1.5 to 7
MSE	Battalion to Corps	64[d]	1.7
MSE with ATM Switch	Brigade to Corps	2,048	5.1 to 6.7[e]
DSCS-111/93	Division to Army	256	27[e]
DSCS-111/85	Division to Army	768	82[e]
SMART-T	Brigade to Army	4,620	481[e]
STAR-T[f]	Corps to NCA[g]	24,000	2,500[e]

Source: Congressional Budget Office based on the Army's 1999 budget hearing for its command, control, communications, and computer (C4) systems.

a. See the glossary of abbreviations on page 45.
b. These averages are lower than the maximum engineering throughputs because of the bandwidth required for context bits and channelization. The averages apply until about 2007, when the Army will begin to field the initial examples of a new generation of communications equipment (see Chapter 2 and Appendix A for more details). After a transition period between 2007 and 2010, Objective Force units in 2010 are scheduled to be the first units to incorporate the new equipment in its entirety.
c. Used for multiplexers, modems, routers, switches, radio access units, and other equipment.
d. The maximum potential rate is 8.192 megabits per second using the (high-capacity line-of-sight) HCLOS radio—provided the interfaces are programmed to operate at the higher rates and frequencies are available. However, at present, the interfaces are not so programmed.
e. Extrapolated from the reductions in bandwidth that occur for lower-frequency radios.
f. The termination in 2001 (for default, as a result of delays and cost overruns) of the ongoing contract for the STAR-T has cast doubt on the program's future. To fill the void produced by the termination, the Army is currently using commercially available systems of approximately the required throughputs and considering replacement candidates.
g. The NCA, or National Command Authority, refers to the command chain that extends to the Secretary of Defense and the President.

munications and its line-of-sight (LOS) radio communications. The bandwidth provided by BLOS satellite communications is several orders of magnitude greater than that available to battlefield LOS radios. The brigade can communicate with higher commands using the BLOS transmitters, as signified by the up arrow. However, it must talk to lower commands with LOS NTDRs, indicated by the down-arrow, because digitized battalions do not have satellite receivers. That satellite communications distinction also divides the *upper tactical Internet* (which is available to commands at the brigade level and above) from the *lower tactical Internet* (available to the brigade and lower command levels).

Bandwidth Demand at Army Commands Today

Experiments have shown that message traffic and data that provide situation awareness and support for decisionmaking, including digital telephone calls, are the bulk of the information transmitted in the Army's communications system. Communications traffic can be thought of as either approximately continuous or approximately episodic. In the former case, called *continuous-flow information*, bits per second is the relevant measure; in the latter case, referred to here as *episodic information*, the size of the message file (in bits) is the appropriate gauge. For instance, situation assessments of U.S. forces that are

Box 3.
Operational Versus Theoretical Bandwidth

Operational point-to-point throughputs are always significantly lower than their theoretical maximums. That reduction stems from two factors: *context bits* and *channelization*.

Context Bits

For military radios, the requirement for context bits for digital throughput is an amalgam of a number of elements. Most of them are simply the result of trade-offs among the engineering constraints associated with hardware, software, and waveforms. Conceptually, however, most of the reduction in throughput attributable to context bits derives from the following requirements and considerations:

- *Network Management.* The Army's tactical Internet uses standard commercial Internet formats such as http, TCP, and IP.[1] Each format has a somewhat different overhead, data-packing format, and packet-transmission protocol.

- *Network Architecture.* Particularly at higher headquarters levels, a network's performance in terms of throughput depends on the network's topology—the number of nodes, the router configurations, the numbers and linkages of relays, and the degree to which messages are multicast (sent to multiple nodes simultaneously). In practice, the network's performance degrades as the number of internal nodes, relays, and routers increases.

- *Forward Error Correction and Retransmission.* Overhead (for example, parity bits), checksum generation, and the time spent by the packet transmitter and receiver in checking errors cut bandwidth even when error rates are low. When error rates are high, message retransmissions also reduce effective bandwidth. Forward error correction, which in its simplest form involves parity bits and some form of data redundancy, allows many errors to be fixed without initiating retransmissions. However, when forward error correction fails, packet retransmissions may be initiated by both the transmitter and the receiver. Although often transparent to the communications system's user, packet retransmission requests and packet retransmissions often dominate message traffic at longer ranges where "noise" becomes a significant problem.

- *Encryption and Decryption.* Throughput can be slowed if encryption/decryption devices do not operate fast enough to deal with the message traffic.

- *Frequency-Hopping Overhead.* Radios can be designed to randomly change the frequencies at which they are transmitting to make it harder to detect and jam transmitted signals. The processing associated with frequency hopping reduces effective throughput.

- *Time-Hopping Overhead.* Radios can be designed to randomly stagger their transmissions to make signal detection and jamming more difficult. The processing associated with those delays reduces effective throughput.

- *Software Inefficiencies.* Hardware devices and the software that makes them function are designed according to certain assumptions about the communications network in which they will operate and the rates of information flow that will apply. As networks age (that is, become loaded with more and newer devices), those assumptions about the network configuration may become invalid, which can lead to reduced throughput.

Channelization

Channelization is the term used to express the number of different data flows a radio can simultaneously support. Simpler radios like the Single-Channel Ground and Airborne Radio System will support either one data channel or one voice channel but not both at once. (The operator flips a switch to use one or the other.) The mobile subscriber exchange—the Army's digital phone system—can be manually set to simultaneously handle large numbers of phone calls (high channelization) or reconfigured to provide one higher-bandwidth channel (low channelization). Low channelization might be used to support a video teleconferencing transmission. The greater the channelization, the lower the effective point-to-point throughput will be.

---

1. Respectively, hypertext transfer protocol, transmission control protocol, and Internet protocol. Each format is optimized differently for different types of data.

Table 2.
Number and Types of Radios Available at the Ops Nodes in Digitized Units During Peak Operations in 2003, by Command Level

Command Level[a]	SINCGARS Links	EPLRS Links	NTDR Links	Satellite Links
Corps	1	1	1	1
Division	1	1	1	1
Brigade	1	1	1	1
Battalion	1	1	1	0
Company	1	1	0	0
Platoon	1	1	0	0
Squad/Vehicle	1	0	0	0

Source: Congressional Budget Office based on Department of the Army, *Army Tactical Communications* (1999), which details the architectural design of the digitized Army.

Notes: In addition to the radios, or "pipes," noted above, which form the trunk lines for the flow of digital information, operations officers may sometimes use analog voice-only systems such as walkie-talkies or several types of purely analog radios. Those analog systems carry a communications load that is not significant for CBO's analysis.

The numbers represent the average number of pipes allocated to the operations officer. For instance, in the column labeled "Satellite Links," a digitized brigade command center usually has three independent satellite terminals (and satellites) available. However, during peak operational conditions, intelligence officers and fire-support activities are typically using two of the terminals—leaving only one, on average, for the operations officer.

See the glossary for abbreviations.

a. At the higher command levels, the table refers to the operations networks only. At lower levels, the distinctions between the various communications networks (for example, operations, intelligence, and fire-support) become less clear.

---

arriving asynchronously every minute or so from a number of points are best represented as a continuous flow of information. By contrast, a command for a *dynamic retasking* (also known as a *dynamic unit task order*, or UTO) may occur only once a day.[6]

Bandwidth Demand for Continuous-Flow Information

The Army has studied the impact of digitization on peak continuous bandwidth demand at the division, brigade, and battalion levels. (For details of that study, see Appendix B.) For continuous-flow information, the digitized division generates a peak demand on the operations net of between 2.5 Mbps and 4 Mbps. Major consumers of bandwidth are ordinary telephone traffic, the Army Battle Command System (ABCS), and video teleconferencing (*see Table 4*).[7] The ABCS is differentiated into segments for classified and unclassified information.

Although data on bandwidth demand are available for the brigade and battalion levels of command, similar data do not exist for the other command levels. Nonetheless, peak continuous demand for them can be extrapolated from the data available for the three higher command levels. The key assumption in the extrapolation is that infor-

---

6. A requirement for a dynamic retasking occurs when a unit is "chopped" from one commander to another—for instance, when a battalion that is subordinate to one brigade at the start of battle is reassigned to another brigade during the course of an operation. Many command and logistics responsibilities are severed and reestablished in such a retasking, requiring the reconfiguration of many network files (for example, address lists and control files).

7. The ABCS is a so-called system-of-systems created from the Army's Tactical Command and Control System (ATCCS) and the Force XXI Battle Command, Brigade and Below (FBCB2) program. The ATCCS is one of the original command-and-control system-of-systems. Created in 1979, it serves as an organizational umbrella for several Army command-and-control programs such as the All Source Analysis System and the Army Field Artillery Tactical Data System, which were initiated in 1969 to manage intelligence and artillery data, respectively. FBCB2 is a system of hardware, software, and databases providing automated capability for situation awareness and command and control to brigade and lower command levels.

mation requirements change from command level to command level in proportion to alterations in the span of control over resources and personnel. Since resources and personnel increase by roughly a factor of three at each command level, so, too, will information requirements. In its estimates of bandwidth demand for continuous-flow data, therefore, CBO assumed that at the corps level, the demand for bandwidth was three times the demand at the division level and that each command below the battalion level required one-third less bandwidth than the next higher command.

That "factor-of-three" assumption has been analyzed both in this study and by the Army (see Appendix B for details). Data on information flows at several different levels of command, although incomplete, were analyzed in one of the Army studies surveyed for this report; the results suggest that the actual scaling factor probably lies between two and four.[8] However, for the sake of specificity and because the results of the analysis remain qualitatively the same, CBO used a factor of three in its analysis. (*Table 5* presents the results of applying the factor-of-three extrapolation to the data in Table 4.)

Table 3.
Maximum Effective Bandwidth Available to Army Operations Networks in 2003, by Command Level

(In kilobits per second)

Command Level[a]	Total Effective Throughput[b]
Corps	2,550
Division	533
Brigade[c]	533↑
	37↓
Battalion	37
Company	15
Platoon	15
Squad/Vehicle	1.7

Source: Congressional Budget Office.

a. At the higher command levels, the table refers to the operations networks only. At lower levels, the distinctions between the various communications networks (for example, operations, intelligence, and fire-support) become less clear.

b. Point to point, under an assumption of perfect load balancing. In practice, throughput rates may be less.

c. The up-arrow (↑) indicates the throughput rate for communications to equivalent or higher command levels. The down-arrow (↓) indicates the throughput rate to lower command levels.

Table 4.
Peak Demand for Continuous-Flow Bandwidth in the Digitized Division's Operations Net in 2003

(In kilobits per second)

Source of Demand	Size of Demand
Army Battle Command System	
Classified, nonlogistics traffic	300 to 1,000
Unclassified, mostly logistics traffic	100 to 300
Telephone (Digital)[a]	1,400
Unmanned Aerial Vehicles	100 to 300
Video Teleconferencing	1,000

Source: Congressional Budget Office based on data from the Department of the Army (see Appendix B for details).

a. Analog telephone communications employing a number of analog radio and walkie-talkie systems are also used but are not considered in this analysis because they are either independent of the communications trunk lines or contribute little to throughput demand.

Bandwidth Demand for Episodic Data

Episodic transmissions on the battlefield fall into several categories: fragmentary orders (FRAGOs), operations orders (OPORDs), fused intelligence reports, UTOs, and map resynchronizations.[9] As noted earlier,

---

8. Barsoum, "Bandwidth Analysis (ACUS Only)." The data and analysis are discussed in Appendix B.

9. FRAGOs are usually short amendments to operational orders and are transmitted as character data. OPORDs are rather lengthy operational orders that are issued once or twice daily. They comprise character-based information including force subordination, disposition of the enemy, the commander's intent, operational goals for the commander's force elements, and decision branch points. Fused intelligence reports are the intelligence summaries produced by intelligence officers that are shared with operations officers. Map resynchronizations, or "resyncs," occur because information associated with digital maps must be updated when a unit moves from one map sector to another.

Table 5.
Peak Demand for Continuous-Flow Bandwidth in 2003, by Command Level

(In kilobits per second)

Command Level[a]	Army Battle Command System Updates		Digital Telephone	VTC[b]	UAV and Common Sensors[b]
	Classified	Unclassified			
Corps	1,000 to 3,000	300 to 1,000	4,000	1,000	100 to 300
Division	300 to 1,000	100 to 300	1,400	1,000	100 to 300
Brigade	100 to 300	30 to 100	300	300[c]	100 to 300
Battalion	30 to 100	n.a.	100	300[c]	100 to 300
Company	10 to 30	n.a.	30	n.a.	n.a.
Platoon	3 to 10	n.a.	10	n.a.	n.a.
Squad/Vehicle	1 to 3	n.a.	3	n.a.	n.a.

Source: Congressional Budget Office based on data from the Department of the Army (see Appendix B for details).

Notes: The estimates of bandwidth demand in this table were calculated by applying a "factor-of-three" extrapolation to the data in Table 4.

UAV = unmanned aerial vehicle; n.a. = not applicable for that command level.

a. At the higher command levels, the table refers to the operations networks only. At lower levels, the distinctions between the various communications networks (for example, operations, intelligence, and fire-support) become less clear.
b. Video teleconferencing and UAVs have bandwidth requirements that are independent of command level.
c. In the digitized battalion, collaborative planning, which is a component of the classified Army Battle Command System, and video teleconferencing are not in operation simultaneously.

the relevant measure of such messages is the size of the file being sent, which CBO estimated on the basis of data gleaned from the advanced warfighting experiments (AWEs) held between 1998 and 2001 at the brigade and division levels and from subject matter experts in communications. Collectively, those experts, who were drawn mainly from the operations, C4 (command, control, communications, and computers), and resources offices of Army headquarters, have experience at multiple command levels generating and using such types of episodic messages (see Table 6).

Calculating peak demand for continuous-flow data at all command levels requires an estimate of the average frequency with which episodic throughputs occur. But no such data have been collected. Therefore, CBO has again generated assumptions, based on information from Army subject matter experts, about such frequencies. FRAGOs and fused intelligence reports are assumed to arrive once a minute at commands below the brigade level and once an hour at commands above it. OPORDS, UTOs, and map resynchronizations are assumed not to occur during peak operations.

Using those assumptions, CBO calculated the equivalent peak continuous information flow associated with those episodic transmissions. Bandwidth demand for those messages, CBO estimates, would be significant only at the lowest levels of command (see Table 7 on page 14). CBO then combined the estimates for both continuous-flow and episodic transmissions to yield an estimate of total continuous bandwidth demand at each command level (see Table 8 on page 15).

Comparing Bandwidth Supply and Demand in 2003

CBO compared the maximum effective bandwidth available at the operations desks of various tactical command levels (shown in Table 3) with its estimates of total demand at those levels (from Table 8). The results of CBO's comparison show that at no command level is there currently a substantial excess of supply relative

Table 6.
File Sizes of Episodic Throughputs in Army Ops Channels in 2003, by Type of Throughput and Command Level

(In kilobits)

Command Level[a]	Fragmentary Orders[b]	Operational Orders[c]	Fused Intelligence Reports	Unit Task Orders	Map Resyncs[d]
Corps	100	10,000	100	1,000	3,000 to 10,000
Division	100	1,000	10 to 100	1,000	1,000 to 3,000
Brigade	10	100	10 to 100	1,000	300 to 1,000
Battalion	1	10	10 to 100	1,000	100 to 300
Company	1	10	10 to 100	100	30 to 100
Platoon	1	10	10 to 100	100	10 to 30
Squad/Vehicle	1	10	10 to 100	100	3 to 10

Source: Congressional Budget Office based on data from Army subject matter experts and from the advanced warfighting experiments held between 1998 and 2001 at the brigade and division levels.

a. At the higher command levels, the table refers to the operations networks only. At lower levels, the distinctions between the various communications networks (for example, operations, intelligence, and fire-support) become less clear.
b. Usually short amendments to operational orders, which are transmitted as character data.
c. Issued once or twice daily, operational orders comprise character-based information including force subordination, disposition of the enemy, the commander's intent, operational goals for the commander's force elements, and decision branch points.
d. Map resynchronizations, or "resyncs," occur because information associated with digital maps must be updated when a unit moves from one map sector to another.

to demand. In the worst case (for communications from the brigade to lower command levels), demand exceeds supply by an order of magnitude or more (that is, by more than 10 to one). In the best case (at the platoon level), demand is somewhere between one-half and twice the bandwidth supply.

To strengthen the qualitative aspects of CBO's comparisons, color coding has been used in *Table 9 on page 16*. At one extreme—for communications from the brigade ops net to operations desks at lower levels and for the battalion ops net—red implies that demand exceeds supply by a factor of 10 or more. A cautionary yellow reflects a match between supply and demand to within a factor of three. (Caution is warranted because although, on average, messages can be expected to get through the network on time, some delays in transmission may be experienced.) Shades of orange indicate gradually worsening imbalances between supply and demand, lying between the cautionary yellow range and the substantial mismatch indicated by red. (In the event of results showing that supply exceeds demand by at least a factor of three, green will be used.)

How does the Army's experience compare with CBO's results? The Army has carried out relatively little testing of large operations nets. However, when it has, participants have invariably cited shortfalls in bandwidth supply as a significant problem. The Army's advanced warfighting experiments conducted at the National Training Center in 1997 and 1998 were, respectively, battalion- and brigade-level experiments using state-of-the-art communications equipment. The AWEs revealed bandwidth problems and network failures to the point where soldiers switched back to analog voice communications as the transmission of digital data slowed. During the Division Capstone Exercise, which was undertaken in April 2001 at the culmination of the four-year effort to develop the digitized division, computer crashes occurred that were attributed to overloaded communications systems.

Table 7.
Equivalent Peak Continuous-Flow Bandwidth for Episodic Throughputs in 2003, by Throughput Type and Command Level
(In kilobits per second)

Command Level[a]	Fragmentary Orders[b]	Operational Orders[c]	Fused Intelligence Reports	Dynamic Unit Task Orders	Map Resyncs[d]
Corps	0.03	0	0.03	0	0
Division	0.03	0	0.003 to 0.03	0	0
Brigade	0.003	0	0.003 to 0.03	0	0
Battalion	0.017	0	0.17 to 1.7	0	0
Company	0.017	0	0.17 to 1.7	0	0
Platoon	0.017	0	0.17 to 1.7	0	0
Squad/Vehicle	0.017	0	0.17 to 1.7	0	0

Source: Congressional Budget Office based on data from Army subject matter experts and from the advanced warfighting experiments held between 1998 and 2001 at the brigade and division levels.

a. At the higher command levels, the table refers to the operations networks only. At lower levels, the distinctions between the various communications networks (for example, operations, intelligence, and fire-support) become less clear.
b. Usually short amendments to operational orders, which are transmitted as character data.
c. Issued once or twice daily, operational orders comprise character-based information including force subordination, disposition of the enemy, the commander's intent, operational goals for the commander's force elements, and decision branch points.
d. Map resynchronizations, or "resyncs," occur because information associated with digital maps must be updated when a unit moves from one map sector to another.

Some insight is also available from more limited tests of networks in the development stage. For instance, at Ft. Hauchuca in 1999, a company-level test evaluated a limited communications network equipped with the Army's most advanced communications gear. Severe degradation occurred in the rate of message completion—it dropped to less than 60 percent—as the network's message load was increased to peak operational levels. In particular, for Blue situation assessments that were successfully transmitted, the latency, or delay, averaged four minutes—a rate that may compromise certain missions.

Long latencies—delays in message transmission or receipt—are often observed in the context of exercises. However, shortages of bandwidth may not only produce latencies in communications networks but in some cases also exacerbate other causes of delays. Operations officers remark that although long latencies are not a handicap for deliberate planning or low-intensity operations, they preclude reliance on the network "when the shooting starts."

Army studies have attempted to estimate the performance of operations networks for large units in the field. One of the studies hints at the existence of the bandwidth bottleneck that CBO's analysis projects for communications from the brigade to the battalion:

> "Our analysis revealed that when large secondary imagery dissemination products are provided down to battalion level, the maximum data rate capability of the radio [NTDR] is utilized. This leaves no reserve capacity to the battalion and confirms the need for more data bandwidth between the brigades and battalions."[10]

Operation Iraqi Freedom

The Army's plans for the recent conflict in Iraq included the use of at least one digitized unit in the early

---

10. Army Signal Center and Fort Gordon, Directorate of Combat Developments, *First Digitized Forces System Architecture*, p. 1.

Table 8.
Total Peak Demand for Effective Bandwidth in 2003, by Command Level
(In kilobits per second)

Command Level[a]	Peak Bandwidth Demand
Corps	3,000 to 10,000
Division	2,500 to 4,000
Brigade	800 to 1,300
Battalion	500 to 750
Company	30 to 100
Platoon	10 to 30
Squad/Vehicle	3 to 10

Source: Congressional Budget Office.

Note: Demand is extrapolated from the division level.

a. At the higher command levels, the table refers to the operations networks only. At lower levels, the distinctions between the various communications networks (for example, operations, intelligence, and fire-support) become less clear.

phase. The 4th Infantry Division, a unit of the 5th Corps, employs the communications architecture described in this study. The division was originally slated to deploy through Turkey and then be involved in the early portions of the conflict. Instead, when it finally disembarked in Kuwait, the conflict was deemed to have largely ended. Therefore, even a qualitative comparison of the results of CBO's analysis and the division's experience with bandwidth supply and demand in combat is impossible.

Yet reports indicate that bandwidth was an issue for those units that were engaged in the conflict. In the months prior to the war and in anticipation of operating in combat with the 4th Infantry Division, other units in the 5th Corps, selected units in the Marine Corps's 1st Marine Expeditionary Force, and the British armored division were outfitted with 1,000 equipment sets for interfacing with the 4th Infantry Division. Contractor support was also provided to enable and sustain communications. The resulting system is now called Blue Force Tracker.

The new system did not fully "digitize" any of those other units, however. Instead, their degree of digitization was termed "digitization lite" by Army officers assigned to quickly provide the interfacing equipment. Typical sets, some of which went to forces at the company level, comprised one terminal from the Force XXI Battle Command, Brigade and Below (FBCB2) program, a Global Command and Control System (GCCS) terminal at the higher command levels, an ABCS terminal, and a commercial L-band satellite transceiver (now increasingly common among U.S. trucking firms), together with interfacing gear.[11] By using commercial L-band satellites (the military has not invested heavily in the L-band), widely dispersed commands could keep track of forces that had the GCCS and FBCB2 displays; at the higher command levels, portions of the GCCS supported planning and decisionmaking. (However, the Army could not fully utilize the ABCS because of unmet National Security Agency certification requirements and other inadequacies.)

When not challenged by line-of-sight constraints, soldiers used their SINCGARS and EPLRS equipment for voice communications. When the widely dispersed and rapidly moving forces faced LOS problems, soldiers substituted commercial e-mail and "chat room" messages for point-to-point and collective voice communications. The mobile subscriber exchange proved useful between fixed communications sites but much of the time was not mobile enough for troops on the move.

At the tactical operations centers at the higher command levels, the minimal equipment configuration described above supplied digital bandwidth that was about one-quarter to one-third of that available in fully digitized units. On the demand side, there was no basis for comparison, for several reasons: during the conflict, trunk lines were often "saturated"—all available digital bandwidth was used up; demand was not subject to the monitoring afforded in tests; lower command levels were not equipped with the FBCB2; and the full capabilities of the ATCCS could not be exploited.

---

11. The GCCS comprises both hardware and applications-level software on the network.

Table 9.
Effective Bandwidth Supply Versus Peak Demand in 2003, by Command Level
(In kilobits per second)

Command Level[a]	Bandwidth Supply	Peak Bandwidth Demand	Relative Supply Versus Peak Demand (S : D)[b]
Corps	2,550	3,000 to 10,000	1 : 1 to 4
Division	533	2,500 to 4,000	1 : 5 to 8
Brigade[c]	533↑	800 to 1,300	1 : 1.5 to 3↑
	37↓		1 : 20 to 30↓
Battalion	37	500 to 750	1 : 10 to 20
Company	15	30 to 100	1 : 2 to 6
Platoon	15	10 to 30	1 : 0.5 to 2
Squad/Vehicle	1.7	3 to 10	1 : 2 to 6

Source: Congressional Budget Office.

a. At the higher command levels, the table refers to the operations networks only. At lower levels, the distinctions between the various communications networks (for example, operations, intelligence, and fire-support) become less clear.

b. Based on an approximate logarithmic scale, the color coding is as follows: yellow indicates that supply is between about one-third and three times demand (a marginal supply/demand match); light orange signifies that demand is approximately three times supply and orange, that demand is approximately three to 10 times supply. Red (used here for the lower brigade-level relationship and at the battalion level) means that demand exceeds supply by a factor of 10 or more.

c. The up-arrow (↑) indicates the throughput rate for communications to equivalent or higher command levels. The down-arrow (↓) indicates the throughput rate to lower command levels.

Despite those drawbacks, some insights might be gained. Fully coordinated Department of Defense "official lessons" from Operation Iraqi Freedom are not yet available, and quantitative data may never be. But qualitative statements from some authoritative sources have been issued.[12] Limited bandwidth was a constant problem in spite of the large increase relative to that available during Desert Storm; bandwidth from commercial satellites in particular was much more heavily used.[13] Although the Army has made a substantial investment in military-only decision-support systems, much of the planning and collective decisionmaking that occurred during the Iraq war was handled through commercial e-mail and chat-room applications that soldiers were familiar with, that were "user friendly" and reliable over long distances, and that required little or no training. Another factor driving forces to use chat-rooms and e-mail was that the distances over which messages had to be transmitted precluded the use of LOS radios and rather than wait or hope for reception to improve (as a result of more relays being deployed), they turned to satellite communications to speed their operations.

Blue Force Tracker was praised by most users in both the Army and the Marine Corps.[14] However, the rapid

---

12. The following discussion draws heavily on the after-action reports (AARs) issued by major units engaged in the conflict: those of the 5th Corps (entitled "V Corps: C4ISR Integration AAR"), the 3rd Infantry Division (see Chapters 17 and 26), and the 2nd Brigade, 101st Airborne Division.

13. At the peak of the conflict, the Defense Information Systems Agency claimed that 3 Gbps of satellite bandwidth was being provided to the theater, 84 percent of which was commercial. That amount is 30 times the satellite bandwidth made available during Desert Storm. See "DISA Chief Outlines Wartime Successes," *Federal Computer Week*, June 6, 2003.

14. One senior Marine Corps officer told CBO that while arguments might continue over the lessons learned from Operation Iraqi Freedom, there seemed to be community-wide agreement on Blue Force Tracker. To paraphrase, "We used to lose units all the time. Subject to bandwidth constraints, with Blue Force Tracker we could find them almost immediately. . . . We're going to buy more."

pace of much of the operation and the wide dispersal of U.S. forces made LOS communications challenging, which, as noted earlier, rely on equally rapid and widely dispersed relays to remain interconnected. (As one example, shortfalls of bandwidth for such communications were nevertheless noted as "information overload," despite the limited information flow those applications—mainly e-mail and chat rooms—generated and the relatively small number of network users.)[15] The formats provided by Blue Force Tracker ranged from bandwidth-intensive imagery to low-bandwidth e-mail text messages. But because bandwidth saturation affected communications for both the Army and Marine Corps, users were often forced to employ message formats that used the least possible bandwidth (in other words, text messages).

Other observations during the operation were related to the availability and survivability of the bandwidth supply. Battery deficiencies (electrical generators are too slow to deploy with rapidly moving forces) were noted, particularly at executive command levels, and concerns were raised about the communications systems' electronic "footprint."[16]

---

15. J. Davis, a reporter embedded with the Army's 11th Signal Brigade in southern Iraq, reported in the June 2003 issue of *Wired Magazine* that "[b]ecause anyone on Siprnet [a DoD classified network] who wanted to could set up a chat, 50 [chat] rooms sprang up. ... The result: information overload." In the same article, Lt. Col. N. Mims of the 11th Brigade is quoted as saying, "We've started throwing people out of the rooms who don't belong there."

16. An electronic footprint is the size, or power, of electromagnetic emissions at various distances. J. Burias, in "G-6 Says OIF [Operation Iraqi Freedom] Validates IT Transformation Path," *Army Link News*, May 30, 2003, quoted Lt. Gen. P. Cuviello: "Antenna farms sprang up around major Army units in both Afghanistan and Iraq as different antennas were needed for each of six different satellite bands and four different types of radios. ... All those antennas sometimes caused co-site interference with each other." Mobility requirements in Iraq and Afghanistan forced troops to rely heavily on batteries rather than the generators normally used at fixed locations. In the same article, Cuviello is also quoted as saying, "Batteries are heavy items to carry around the battlefield—not only to keep them stocked and transported, but also the transportation requirements to dispose of them."

CHAPTER 2

Bandwidth Supply and Demand in 2010

The first units of the Army's future Objective Force are expected to be operational in 2010. Those units will employ new, high-bandwidth communications systems that provide about an order of magnitude more bandwidth than that available today. However, at certain levels of command, the Congressional Budget Office projects, the growth of bandwidth demand will outstrip the growth of supply. Changes in demand are expected from three sources: incremental growth associated with upgrades to currently deployed systems; more-substantial growth from the introduction of new systems that have much greater capacity to process and generate information; and changes in the ways that information is processed and exchanged.

Bandwidth Supply at Army Commands in 2010

Three major programs are expected to increase the Army's supply of bandwidth by 2010. The Joint Tactical Radio System (JTRS) is being designed to boost the supply of bandwidth for the lower tactical Internet. The Warfighters Information Network-Tactical (WIN-T) and a new satellite terminal program, which is expected to provide the capabilities of the canceled STAR-T program, are planned to increase bandwidth supply for the upper tactical Internet.[1]

The Joint Tactical Radio System

The JTRS program is a complicated joint effort involving the four military services and the Special Operations Command. By 2010, the Army-led portion of it will be fielding its enhanced radios across the command chain. The JTRS will replace the Army's current SINCGARS, EPLRS, and NTDR equipment with higher-capacity, multichannel software-defined radios (SDRs). Most digital radios in the field today are not SDRs; instead, they produce their signals through their hardware alone and consequently lack much of the flexibility of SDRs.[2]

The JTRS will be capable of communicating with the Army's "legacy" radios (the current generation of equipment) and will use a wide-band network waveform (WNW) to provide high-capacity bandwidth.[3] The Army's

---

1. The ongoing STAR-T (SHF [Super High Frequency] Triband Advanced Range Extension Terminal) contract was terminated in 2001 for default, as a result of delays and cost overruns.

2. Appendix A provides more detail about the differences between SDRs and traditional digital radios. "Flexibility" pertains to the relative cost, in time and money, of modifying the radio frequency waveform (see the footnote below) in the entire "fleet" of fielded radios. In principle, SDRs should be cheaper to modify because the modifications involve changing only their software, which can be done in the field, rather than pulling and replacing circuit boards in depots.

3. Every radio has at least one waveform. Waveforms are distinguished by the set of engineering choices made about such aspects as frequency; amplitude modulations (as in an AM radio) or phase modulations (FM); framing (breaking the data-carrying sections into blocks); frequency hopping; time hopping; and a number of other elements. Many of those other elements are designed to improve the reliability of transmissions to intended recipients, yet decrease the probability of detection and interception by others.

Table 10.
Effective Bandwidth Available to Army Operations Networks in 2010, by Command Level

(In kilobits per second)

Command Level[a]	Total Effective Throughput[b]
Corps	3,100
Division	1,100
Brigade[c]	1,100 ↑
Battalion	600 ↓
	600
Company	400
Platoon	400
Squad/Vehicle	200

Source: Congressional Budget Office.

Note: In developing its estimates, CBO assumed that no change would be made in information management or network architecture.

a. At the higher command levels, the table refers to the operations networks only. At lower levels, the distinctions between the various communications networks (for example, operations, intelligence, and fire-support) become less clear.

b. Point to point, under an assumption of perfect load balancing. In practice, throughput rates can be expected to be lower.

c. The up-arrow (↑) indicates the throughput rate for communications to equivalent or higher command levels. The down-arrow (↓) indicates the throughput rate to lower command levels.

goal for the WNW is an engineering bandwidth of about 2 megabits per second, which would give an operational point-to-point throughput of about 200 kilobits per second. That amount is generally about an order of magnitude more bandwidth than is now provided by any of the radios in use in the lower tactical Internet.

The WIN-T Program and Satellite Upgrades

According to the Army's current plans, the WIN-T program will not only purchase radios but also supply computer terminals and servers, local area networks and the associated networking gear, cryptological devices, other interfacing equipment, and the software to run all of those items. The WIN-T will be fielded at the brigade, division, and corps command levels, interfacing with the satellite terminals that are used by those commands.

To take advantage of the bandwidth provided by the WIN-T's high-capacity radios, the Army plans to upgrade the associated satellite terminals as well. The Multiband Integrated Satellite Terminal, or MIST, program will provide those improved satellite communications. Coupling the WIN-T equipment and software to the new satellite terminal will deliver a maximum engineering throughput of about 24 Mbps and effective bandwidth of about 2.5 Mbps.

Total Bandwidth Supply in 2010

Using assumptions analogous to those it adopted to derive the maximum operational bandwidth of the Army's current networks (*see Table 3 on page 11*), CBO has estimated the bandwidth that the JTRS, WIN-T, and MIST programs will provide at all levels of command (*see Table 10*).

Bandwidth Demand at Army Commands in 2010

Several organizations have recently published projections of expected incremental growth in the demand for communications capability on the battlefield.[4] Those projections, which are based on experience over the past several years, indicate annual growth rates that vary from 10 percent to 46 percent, depending on the level of command and other factors. Absent major new program initiatives, the projections imply that the demand for bandwidth can be expected to double, on average, every two to five years across all levels of command. Therefore, to estimate the demand for bandwidth in 2010, CBO adopted the more conservative end of that projection—in other words, that demand

---

4. See Office of the Undersecretary of Defense for Acquisition, Technology, and Logistics, *Report of the Defense Science Board Task Force on Tactical Battlefield Communications* (February 2000), p. 49; Army Signal Center and Fort Gordon, Directorate of Combat Developments, Modeling and Simulation Branch, Architecture Division, *First Digitized Forces System Architecture (1DFSA): Version 2.02, Simulation Analysis/Study*, White Paper (main text and attachment entitled "Satellite Communications Capacity Study," June 30, 1999), p. 3 of the attachment; and RAND Arroyo Center, "Future Army Bandwidth Needs—Interim Assessment" (briefing prepared for the G6/Army Chief Information Officer, July 10, 2002). The RAND interim assessment cites projections of bandwidth demand growth by the Defense Advanced Research Program Agency.

Table 11.
Number of Operating UAV Systems, by Command Level

Unit Size	Digitized Units, 2003	Interim Force	Objective Force, 2010
Tactical UAVs			
Corps	0	0	1[a]
Division	0	0[b]	1 to 2[c]
Brigade	1	1	1 to 2[c]
Small UAVs			
Battalion	0	0	1
Company	0	0	1
Platoon/Squad	0	0	1

Source: Congressional Budget Office.

Notes: Unmanned aerial vehicles are managed in systems of two or four. Typically, at the brigade level and above, a system will be organized in a UAV platoon with four UAVs. Systems of small UAVs (a kind of tactical UAV) comprise two vehicles each. The Army's current plans assume that a small-UAV downlink will not be networked beyond the controlling unit.

During military actions, only one UAV per system will typically be flying; however, additional UAVs may serve as data relays, and other, nonflying UAVs may be used as spares. The numbers in the table represent the number of UAVs that are under the direct control of the command and that are simultaneously initiating transmissions.

a. CBO assumes that the Shadow or another extended-range system will probably be used.
b. ACRs (armored cavalry regiments), which are approximately the size of a brigade, would each have two UAVs.
c. Under current plans, the Army expects to use the Shadow system at this level.

---

for bandwidth would grow by about 15 percent a year and would double every five years.

That estimate of 15 percent annual growth excludes the effects that major new program initiatives might have on demand. However, at least one recent initiative associated with the Army's transformation—the service's intended widespread use of unmanned aerial vehicles—will significantly increase the demand for bandwidth over and above the service's historical annual growth rates.[5] Recently, the Army has been experimenting with UAVs in the digitized forces and Stryker brigades, and they have been used in operations in Iraq, Macedonia, Bosnia, and Kosovo. Over the longer term, the Army envisions a family of UAVs, which will be employed by its Objective Force of the future.

The Army calls most of its UAVs tactical UAVs, or TUAVs, because they are controlled by tactical commanders on the battlefield. (A subset of the TUAVs, those with the shortest ranges, are called small UAVs, or SUAVs.) The Army intends to field some type of TUAV in the three higher command levels (see Table 11).[6] Currently, both the digitized forces and the Stryker brigades are equipped with Hunter TUAVs, but the Army intends to replace them soon, substituting the more advanced Shadow system.[7]

---

5. Office of the Secretary of Defense, *Unmanned Aerial Vehicles Roadmap 2000-2025* (April 2001); and Congressional Budget Office, *Options for Enhancing the Department of Defense's Unmanned Aerial Vehicle Programs* (September 1998).

6. UAVs such as the Global Hawk and the Predator, which are operated by the Air Force at the direction of a theater commander, are examples of strategic UAVs. Although the Army is pursuing some limited funding for research and development of strategic UAVs, it has not made a commitment to field them.

7. A report from the Department of Defense's Director of Operational Test and Evaluation recently characterized the Shadow as neither operationally suitable nor operationally effective, on the basis of an initial operational test and evaluation completed in 2002. However, the Shadow was deployed in Iraq during the recent conflict, and as a result, CBO assumes that the problems noted earlier will be resolved and the system will be fielded by 2010.

Operating the Shadow system at the brigade and higher command levels would generate a sizable demand for bandwidth, depending on the degree to which the information those TUAVs collected was shared throughout the battlefield communications network. The Shadow will have three communications channels. One is a large data channel with an engineering throughput of 16 Mbps that, operationally, should deliver from about 1.5 Mbps to 2 Mbps of useful video bandwidth. The other two are redundant command-and-control channels providing 19.6 Kbps of operational throughput. A division will control between four and eight of these TUAV systems (although division commanders will almost certainly make three to six of them subordinate to brigade commanders).

The doctrine underlying use of the Shadows is still evolving, but at this point, the Army wants to share among brigade and higher command levels the information collected from at least four—and possibly as many as eight—of the TUAVs. Currently, data from TUAV downlinks are shared between the operations and intelligence nets of a command.[8] Under the assumptions that there are three brigades per division and that they will be sharing (that is, networking) the information among themselves and also with their division, then information from four to eight TUAVs will be transmitted on each brigade's operations trunk line. If each TUAV requires 1.5 Mbps of bandwidth, each brigade will need from 6 Mbps to 12 Mbps. Divisions typically command those brigades, and a corps commands the divisions. Hence, the demand for TUAV bandwidth for the sharing of such information at those levels will be from three to nine times larger than at the brigade level.

---

8. Currently, all Army UAVs are considered intelligence assets, although their communications downlinks are shared with the operations channels, which can use them for situation assessments. In the future, the traditionally separate operations and intelligence nets could be consolidated, in which case the demand for bandwidth in the operations net would simply be the sum of the two currently independent demands.

Table 12.
Peak Effective Bandwidth Demand in 2010 Under Two Assumptions About TUAVs, by Command Level
(In kilobits per second)

Command Level[a]	Peak Bandwidth Demand	
	TUAVs Are Autonomous	TUAVs Are Networked
Corps[b]	10,000 to 30,000	30,000 to 100,000
Division[b]	3,000 to 10,000	10,000 to 30,000
Brigade[b]	1,000 to 3,000	3,000 to 10,000
Battalion	1,000 to 2,000	1,000 to 2,000
Company	100 to 300	100 to 300
Platoon	30 to 100	30 to 100
Squad/Vehicle	10 to 30	10 to 30

Source: Congressional Budget Office.

Note: TUAV = tactical unmanned aerial vehicle.

a. At the higher command levels, the table refers to the operations networks only. At lower levels, the distinctions between the various communications networks (for example, operations, intelligence, and fire-support) become less clear.

b. If the architecture for information distribution located in the tactical operations centers at these command levels is altered, as the Army is considering doing, then operations officers will share information in a new "backbone net." In that case, the ranges of numbers in the right-hand column (TUAVs are networked) would apply.

The Army's plans for the sharing of TUAV data—with its heightened bandwidth demand—led CBO to develop two different scenarios for total demand in 2010 under the assumption in which the Army's network architecture (broadly speaking, its information management approach, including hardware and software) does not change from its current structure in digitized units. As noted earlier, that structure is one of multiple networks—for example, those for operations, intelligence, and fire support. The information carried in each net can be differentiated by whether or not it was generated by TUAVs.

In the first scenario, existing demand is assumed to grow by 15 percent a year, as discussed previously. The second scenario adds a further assumption: that information collected by the Army's future TUAVs will be shared among the operations networks of the upper command levels. In other words, the first scenario incorporates the assumption that TUAV operations are autonomous (their information is not shared); the second, that their operations are networked (see Table 12).

But the Army is considering altering the network architecture in the future. One such change it is discussing would load the non-TUAV information from all three networks onto a single *backbone net* and divert the burgeoning TUAV information to a new, distinct high-capacity network. In that new architecture, the demand for bandwidth on the backbone net would be approximately three times the demand on the ops net under the old architecture and without networked TUAVs. But the backbone net demand would also be the same as in the case in which TUAVs were networked in the ops net but no architectural change had occurred.

Thus, the most likely estimates of demand for operations bandwidth in 2010 correspond to those that incorporate the assumption that TUAVs are networked, regardless of whether or not the Army changes the architecture of its battlefield information management system.

Comparing Bandwidth Supply and Demand in 2010

As it did for 2003, CBO compared estimates of bandwidth supply and demand at the operations desks of various tactical command levels in 2010. For its demand estimates, CBO assumed that TUAVs would be networked and accommodated in either of the two information management architectures that the Army is considering for 2010 (either the architecture characterized by separate networks to support the ops, intelligence, fire-support, and other missions or the structure featuring a single backbone net for all non-UAV data and a separate, high-capacity network dedicated to transmission of UAV information). The use of those two assumptions regarding architecture is advantageous for two reasons: both assume that TUAVs will be networked (the current strategy), and both impose approximately the same bandwidth demand on the operations network. Under the assumptions that demand grows in accordance with the requirement to accommodate net-

Table 13.
Effective Bandwidth Supply Versus Peak Demand in 2010, by Command Level

Command Level[a]	Relative Supply Versus Peak Demand (S : D)[b]
Corps[c]	1 : 10 to 30
Division[c]	1 : 10 to 30
Brigade[c,d]	1 : 3 to 10↑
	1 : 5 to 15↓
Battalion	1 : 1.5 to 3
Company	1 to 4 : 1
Platoon	4 to 10 : 1
Squad/Vehicle	7 to 20 : 1

Source: Congressional Budget Office.

Note: The range in relative demand at the brigade and higher levels of command is associated with either the proposed "backbone" architecture or the case in which the current architecture is maintained and downlinks from unmanned aerial vehicles are heavily networked.

a. At the higher command levels, the table refers to the operations networks only. At lower levels, the distinctions between the various communications networks (for example, operations, intelligence, and fire-support) become less clear.

b. Based on an approximate logarithmic scale, the color coding is as follows: green (used here at the platoon and squad/vehicle command levels) means that supply exceeds demand by approximately a factor of three of more. Yellow indicates that supply is between about one-third and three times demand (a marginal supply/demand match), and orange signifies that demand is approximately three to 10 times supply. Red (used here for the corps and division levels) means that demand exceeds supply by a factor of 10 or more.

c. If the architecture for information distribution located in the tactical operations centers at these command levels is altered, as the Army is considering doing, then operations officers will share information in a new "backbone net." In that case, the upper end of the projected range of bandwidth demand would apply.

d. The up-arrow (↑) indicates the throughput rate for communications to equivalent or higher command levels. The down-arrow (↓) indicates the throughput rate to lower command levels.

worked TUAVs and that all of the hardware improvements that are now anticipated are fielded, the bandwidth bottleneck will change its location in 2010 from the brigade to the corps level but will be as severe then as it is at the battalion level today (see Table 13).

The degree of mismatch between CBO's supply and demand projections for 2010 varies considerably by command level, as is the case currently. If the JTRS performs as the Army projects it will, the new radio will generally provide more than enough bandwidth for the lower tactical levels of command, including a margin for the potential growth of demand beyond 2010. At the division and corps levels, however, the projected demand swamps the likely supply.

How well do the Army's most recent analyses of future bandwidth supply and demand compare with CBO's estimates? Only partial comparisons are possible, given the limited data. For a brigade-level unit in the future Objective Force under surge conditions (in combat and on the move), Army analysts project a demand for engineering bandwidth of 35 Mbps, which corresponds to an effective demand of about 4 Mbps. The 4 Mbps figure is consistent with the demand that CBO projects at the brigade level of between 3 Mbps and 10 Mbps (see Table 12). Another Army source has indicated that the effective demand could be as high as 12 Mbps—in which case CBO's projection would underestimate the degree of mismatch.[9] The Army concludes that for 2010, the service, "will NEVER have enough BW [bandwidth]" and urges that it be treated as "an operational (limited) resource."[10]

---

9. Personal communication to the Congressional Budget Office by Robert M. Saunders Jr., Deputy Director for Technology (Communications), Office of the Assistant Secretary of the Army for Acquisition, Logistics, and Technology, November 17, 2002.

10. Lt. Gen. P. Cuviello, "Projected Bandwidth Usage and Capacity" (briefing prepared by the G6/Army Chief Information Officer for the Army Chief of Staff, August 2002), p. 4.

CHAPTER

3

Mitigating Mismatches Between Bandwidth Supply and Demand

The Congressional Budget Office has examined several ways to treat the projected mismatches between the Army's demand for and supply of communications bandwidth on the battlefield. Those approaches include increasing supply, decreasing demand, reallocating currently planned spending to focus more resources on command levels with the most severe bandwidth problems, and developing tools to better manage whatever mismatches persist. Increasing the supply of bandwidth by buying more or better radios would probably be the least effective way to mitigate the supply/demand imbalance. Indicators of the efficacy of reallocating resources are ambiguous, and their results depend on the metric being used. CBO is thus unable to provide any unequivocal findings on that option. Decreasing demand would help lessen the mismatch, and some potential options are discussed, although they would not eliminate the imbalance. Better management of supply and demand would also help shrink the supply/demand gap, and several aspects of that approach are noted.

Buy Better Radios in Greater Quantities

The Army might increase its supply of bandwidth in 2010 by buying radios that are more technologically advanced than those it is currently expecting to have. Yet the service's planned programs appear to already be incorporating all of the most likely advances in communications technology. To develop new radios such as the Joint Tactical Radio System, the Army has contracted with large corporate teams that include many of the world's premier radio makers and computer network experts, and program managers report that those development efforts are fully funded.[1] Because of the ambitiousness of some of the technological advances being pursued, there is a risk that not all of them will be realized (*see Appendix A*). Thus, it seems unlikely that advances in technology over and above those that the Army is already pursuing could reduce the bandwidth supply/demand mismatches that CBO projects.

Also questionable is whether the projected shortfalls in the supply of Army bandwidth could be eliminated by purchasing more of the communications equipment that is now being developed. At the division and corps levels, CBO estimates, demand in 2010 will outstrip supply by a factor ranging between 10 and 30. At those levels of command, the Warfighter Information Network-Tactical program will supply the majority of bandwidth, and the Army has considered increasing its purchases of WIN-T equipment 20-fold to better match the expected demand. But implementing such an option would present a number

---

1. CBO analysts questioned program managers about the adequacy of funding levels as one way to gauge the severity of the risk that products scheduled for fielding by 2010 might not actually be available. Managers reported full funding, which signifies the Army's commitment throughout the Department of Defense's Future Years Defense Program to provide funds that the service and the Defense Department agree will meet the programs' formal requirements. For products scheduled for fielding after 2010 (for example, JTRS Cluster 2 cellular phones), program managers identified unfunded requirements that if fully funded might accelerate fielding dates. However, bringing those capabilities to the field faster would not mitigate the shortfalls that CBO's analysis revealed.

Figure 1.
Notional Throughput Capacity per Node
(In kilobits per second)

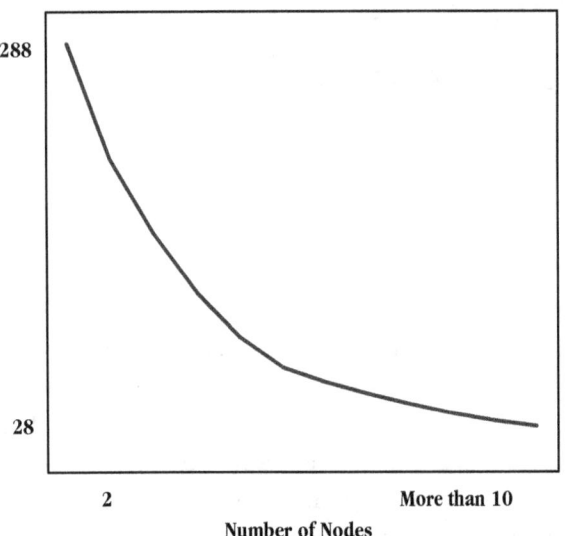

Source: Congressional Budget Office based on a briefing on the upper and lower tactical Internets provided by the Army Chief of Staff, Staff Group.

Note: The rate of throughput is highly dependent on the number of nodes in the network.

of challenges, including management of the increased complexity and inefficiency of the communications network that would result; the physical space, or "footprint," required to accommodate the additional equipment on the battlefield; and the additional cost.

Increased Network Complexity and Costs

As the number of nodes in a communications network increases, the network's efficiency—measured as the throughput capacity available at each node—decreases (see Figure 1). As the figure shows, an order-of-magnitude *increase* in the number of nodes contained in the network can produce nearly an order-of-magnitude *decrease* in bandwidth at a given node. The results in the figure are consistent with theoretical estimates of how a network's efficiency (measured in kilobits per second) varies with the number of nodes when current methods (specifically, Internet processing protocols) for transmitting information around a network are used.

Theory indicates that network throughput can be expected to decay as $1/Q$, where Q is the square root of $N \ln(N)$, N is the number of nodes in the network, and ln is the natural logarithm.[2] Consequently, if the number of radios increased by a factor of 20, the network's throughput would rise by only a factor of 20/7—or about 3—where 7 is the reduction in throughput caused by the increase in the network's complexity.[3] In other words, the 20-fold increase in purchases of WIN-T equipment that the Army has considered would not be sufficient to overcome the 10- to 30-fold shortfall in available bandwidth that CBO projects for 2010 at the corps and division levels of command. Theoretically, at least a 600-fold increase would be required, which the Army is unlikely to pursue, given that WIN-T procurement costs are currently estimated to be about $5 billion.[4]

Expansion of the Footprint

The geographic footprint of a program is the physical space that the program's equipment occupies when it is operated on the battlefield. The Army's current plans for the WIN-T and JTRS have raised concerns about the proliferation of equipment in vehicles and operations centers. Part of the reason for that concern is a lack of physical space, an element that has also surfaced in discussions about changing the network architecture. Another issue is that the associated increase in the equipment's electronic footprint (the size, or power, of its electromagnetic emissions at various distances), caused by the proliferation of

---

2. See P. Gupta and P.R. Kumar, "The Capacity of Wireless Networks," *IEEE Trans Information Theory*, vol. 46, no. 2 (March 2000), pp. 388-404.

3. The development and adoption of new methods—including so-called dynamic protocols—for transmitting information among the nodes in a communications network might reduce the inefficiency penalty that accrues as the number of nodes increases. However, the development and adoption of such protocols for use in military communications networks are unlikely to occur prior to 2010.

4. The Army's estimates in 2002 of the total cost for the WIN-T program ranged from $4 billion to $9 billion (in 2002 dollars), with $2 billion in expenditures scheduled between 2004 and 2009. More recent (2003) estimates by contractors put total program costs at about $5 billion and $6.6 billion, respectively. The figures that follow use $5 billion. (Unless otherwise indicated, all costs are expressed as fiscal year 2003 dollars.)

radiating antennas and electronic "noise," will facilitate enemies' detection and targeting of U.S. forces. In that context, increasing the amount of communications equipment by a factor of 20—or more—could pose significant problems.

Reallocate Currently Planned Spending

If additional spending is unlikely to close the gap between bandwidth supply and demand, might currently planned spending be reallocated to focus more on the command levels that projections show will have the most severe problems in 2010? Answering that question requires constructing so-called figures of merit to relate projected spending at each command level to communications capacity—both the capacity that will be provided and the capacity that will be demanded. Several such metrics can be considered.

Spending data indicate that over the next several years, the Army plans to spend more money at the lower levels

Figure 2.
Total Projected Investment in Ops Net Equipment, by Command Level
(In billions of fiscal year 2003 dollars)

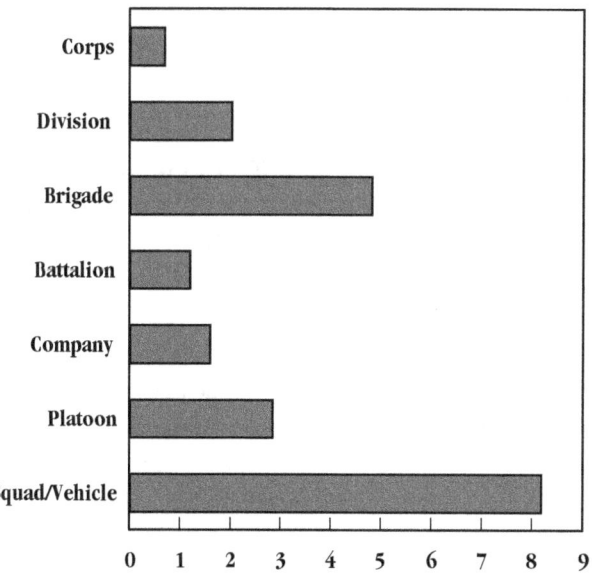

Source: Congressional Budget Office.

Figure 3.
Cost of the Capacity Provided per Ops Net, by Command Level
(In fiscal year 2003 dollars/bit per second/net)

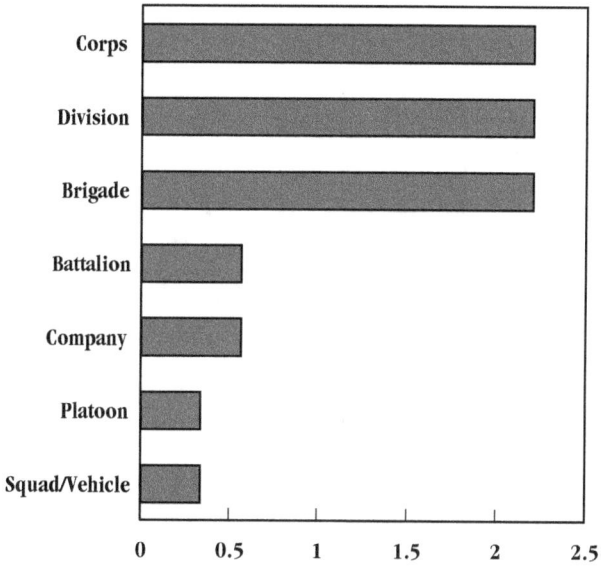

Source: Congressional Budget Office.

of command—that is, below the division and corps levels and away from the bandwidth bottleneck projected for 2010 (see *Figure 2*). That distribution of expenditures arises because radios are spread throughout the force and most personnel with radios are located not at higher command-level headquarters but in the decreasingly smaller, more numerous units that compose the lower levels of command.

Considering the cost of the total bandwidth capability that the Army plans to provide yields a different picture. Lower levels of command comprise many vehicles and soldiers, which the Army envisions equipping with relatively simple versions of the JTRS. At higher command levels, smaller numbers of more complex and more expensive versions of the JTRS will be fielded, as well as more sophisticated—and costly—line-of-sight and beyond-line-of-sight satellite communications radios. That latter equipment will supply much greater amounts of bandwidth per unit and support more networks than the simpler but more numerous radios planned for the lower levels of command.

Figure 4.
Cost of the Capacity Utilized per Ops Net, by Command Level
(In fiscal year 2003 dollars/bit per second/net)

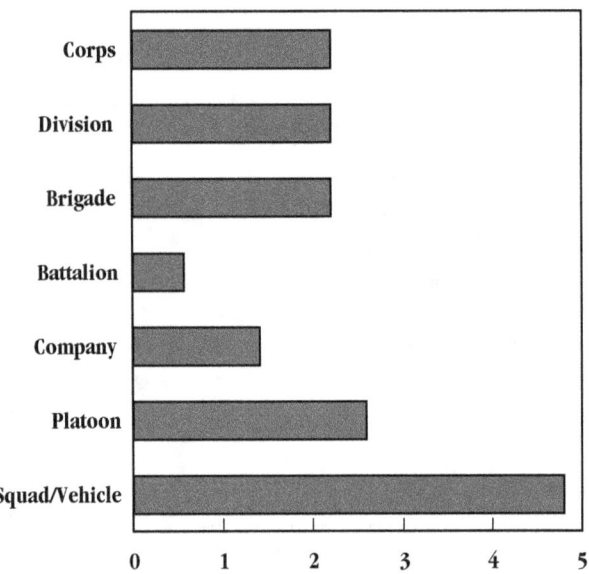

Source: Congressional Budget Office.

A metric can be constructed that accounts for the cost of the capacity provided per ops net at each command level (*see Figure 3*). As the figure shows, the cost is greatest for the levels of command at which the bandwidth bottleneck is projected to be most severe in 2010.

Considering the cost of the capacity utilized (as opposed to provided) yields yet another view of how the Army's projected investments are distributed by command level (*see Figure 4*). As Chapter 2 discussed, the supply of bandwidth at the squad and platoon levels in 2010 will exceed the projected demand by factors ranging from four to 20. The reason is that the simpler version of the JTRS to be fielded at lower command levels is equipped to communicate with higher levels; that is, it has the capability to use the wide-band networking waveform (discussed in Appendix A), which provides substantial bandwidth—more than the demand now being projected. A radio lacking that capability would be less expensive and could allow resources to be channeled elsewhere. However, it would also be incapable of communicating with all levels of command or with the forces of the other military services.

Would a reallocation of the Army's projected resources be a viable option for mitigating the mismatch between bandwidth supply and demand projected in the future Objective Force? The results presented in the figures above do not offer clear guidance. Consequently, such a reallocation does not appear to be an obvious alternative for addressing the Army's future bandwidth bottleneck.

Reduce Demand and Better Manage Persisting Mismatches

Because of the potential problems associated with buying more bandwidth to eliminate the Army's projected bottleneck, CBO considered options to reduce lower-priority bandwidth needs and to better manage the remaining demand. Three such alternatives are discussed below. The first two would end the transmission of information that might be of lesser priority yet would be expected to contribute significantly (either directly or indirectly) to the future supply/demand mismatch at the higher command levels. The third option focuses on adopting software tools that are now entering the commercial market and that could allow better management of the demand for bandwidth when it exceeds supply. In keeping with its mission of providing impartial analysis, CBO makes no recommendations about adopting one or more of these options. Rather, they are meant to illustrate some of the choices that the Army might consider as it attempts to match its expectations regarding digitization to the communications capability it is buying.

Eliminate Video Teleconferencing on the Battlefield

As discussed earlier, video teleconferencing is a bandwidth-intensive activity; the usual teleconferencing site requires about 1,000 Kbps of bandwidth. Video teleconferencing enables commanders at disparate locations to interact as if they were personally present at one location. Participants can hear vocal tones and observe facial expressions and other aspects of behavior that commanders argue are important elements in decisionmaking and planning. In addition, they can observe and comment collectively on complex visual objects—for example, maps depicting the disposition of enemy and friendly forces.

In cases in which video teleconferencing capability does not exist or is unavailable because of a lack of bandwidth, an alternative is voice conferencing, which has been and continues to be used routinely by commanders as a substitute. Although voice conferencing does not allow conferees to collectively view objects or drawings, that capability can be provided by adding a "smart" whiteboard visual display; the digital whiteboard allows conferees to view a rendition (in black and white, not color, which would require more bandwidth) of the object under discussion—typically a dynamically generated drawing, map, or operational plan. (The conclusion that such a capability could be useful was reached by the 5th Corps just prior to the recent Iraq war.)[5] Conferees can not only use the whiteboard as a blackboard but may also utilize electronic features such as the ability to scan documents and communally view files.

Overall, voice conferencing coupled with whiteboard capability generally requires less bandwidth (by about a factor of 10) than video teleconferencing requires. The Army's test community has assessed the relative value, for collaborative planning, of teleconferencing, a drawing or black-and-white projection capability (available with the whiteboard), and video. Its judgment: "The draw and voice features deliver the highest value for collaboration. The draw feature was particularly critical. Video is a good feature but has the least relative value as compared to draw and voice capabilities."[6]

A second military use for video teleconferencing is telemedicine, in which medical personnel consult with doctors at remote locations to bring to bear experience and expertise that may not be available on the battlefield. Telemedicine has been a feature of recent low-intensity operations (such as the U.S. military's peace enforcement mission in Macedonia) during which only limited medical support was available in the theater of operations. But telemedicine's value in other military operations has been questioned. In-theater medical capability is extensive in high-intensity battlefield situations such as Operation Iraqi Freedom. Above the medical aid stations at the battalion level of command, the medical evacuation system delivers patients to combat-support hospitals and to field and general hospitals. Those large facilities have scores of physicians who are trained in battlefield and specialty medicine and who may not have the time or the need to consult extensively with off-site doctors. Hospital facilities are reinforced by an Air Force-run medical evacuation system that can move patients quickly to major medical centers in Europe, Japan, or the United States. After considering all of those factors, the Army has chosen not to provide telemedicine capability at the battalion level of command.

Eliminating video teleconferencing would reduce the current demand for bandwidth by as much as 70 percent at the brigade command level and 15 percent at the corps level, CBO estimates. Less demand at the brigade level would be helpful in reducing the bottleneck there, but the 15 percent drop in demand at the corps level would mitigate little of the bandwidth supply/demand mismatch there, where it is projected to be more severe in 2010. Those same reductions would apply to the demand for bandwidth in 2010 under the scenario in which video data collected by UAVs were not shared over the network. If such data were shared, which is a more likely outcome, then total bandwidth demand would be greater, and the fractional savings associated with eliminating video teleconferencing would be reduced to about 20 percent at the brigade level and 5 percent at the corps level.

Eliminate the Requirement to Support Networked UAVs

As discussed in Chapter 2, the demand for Army bandwidth in 2010 will exceed the projected supply by as much as a factor of 30 at the division and corps levels of command and up to a factor of 10 at the brigade level. Those estimates apply, regardless of the information management scheme that the Army decides to use, if the video data collected by UAVs are shared among operations networks at the upper levels of command. Under this option, the

---

5. After several years of experience with video teleconferencing (VTC), the 5th Corps's commander, Lt. Gen. William Wallace, suggested in 2002 that VTC "was wasteful," and the corps went to war in Iraq with voice conferencing and a rudimentary whiteboard capability. A lack of training and established procedures prevented those tools from being judged fully successful, but the bandwidth advantage they offered (their bandwidth demand is much less than that of video teleconferencing) continues to spur additional experimentation. See Maj. M. Shaaber, Capt. S. Hedberg, and Troy Wesson, *V Corps: C4ISR Integration AAR* (after-action report) (May 2003), pp. 37-38.

6. Lee Offen and Mary E. Stafford, *Assessment Report for the Division XXI Advanced Warfighting Experiment (DAWE)* (United States Army Operational Test and Evaluation Command, January 22, 1998), p. ES-11.

Army would forgo sharing those data beyond the initial sensor-to-shooter downlink. Instead, UAV operators would summarize collected intelligence in a lower-bandwidth message, which could then be more easily transmitted. Commanders would thus receive the results of UAV missions but would not have direct access, in real time, to the video images that the vehicles collected. Eliminating the requirement to share video data would reduce the future demand for bandwidth at the brigade and higher levels of command by a factor of three.

Provide Bandwidth Management Tools at the Applications Layer

If the Army eliminated both video teleconferencing and the requirement to share UAV data, CBO estimates that the service could reduce total bandwidth demand by roughly a factor of three. Nevertheless, such a reduction would be insufficient to bring demand into line with supply. Currently, when the supply of bandwidth is inadequate, operations officers attempt to manage demand manually as they transmit messages. They prioritize transmissions, on the basis of their experience and professional judgment, assigning a sequence to the times at which messages in a queue will be sent. Sometimes, to decrease delays and increase message completion rates, communications personnel in a tactical operations center literally pull the plug temporarily on some equipment in order to have enough bandwidth for the highest-priority messages. As discussed in Chapter 1, however, those efforts to reduce the demand for bandwidth have frequently failed to prevent unacceptable degradation in the performance of the Army's battlefield communications networks during experiments. During the recent war in Iraq as well, constraints on bandwidth forced soldiers to confront their lack of techniques for managing bandwidth.[7]

Communications officers have automated network management tools available to help them monitor a variety of measures of bandwidth demand and thus aid their prioritizing of message traffic. Some of those tools can automatically increase or decrease the bandwidth allocated to given types of messages or other data transmitted over the Army's battlefield network. But there is little or no feedback in such systems; consequently, the software applications that generate the messages and data transmitted over the network cannot automatically detect that the available bandwidth has been increased or decreased. As a result, message queues tend to lengthen once a bandwidth allocation has been reduced because the rate at which the messages are generated does not change. And once messages become old and exceed thresholds for latency (delay), they are eliminated from the network without being transmitted. In more extreme cases, the rates at which information flows between hardware components (which are usually set, by the flip of a physical switch, when the hardware is linked to the network) are exceeded, and messages are truncated or dropped. Neither condition is satisfactory: lost messages may be important, and overflowing queues, a portion of which the network continually tries to retransmit, strain the network's responsiveness and reliability.

This option would expand the capabilities of today's network management tools so that message-producing software applications could sense changes in the available bandwidth and automatically increase or decrease the rate at which messages were generated. That approach is beginning to be used by private-sector firms that develop software for use on the Internet.[8] The cost of developing and testing the software required to implement this kind of an option for the Army is currently unknown, and CBO did not attempt to estimate it.

---

7. Shaaber, Hedberg, and Wesson noted in *V Corps: C4ISR Integration AAR* (p. 5): "The ability to manage bandwidth usage dynamically at the discretion of the commander would [be helpful]."

8. The Defense Information System Agency cites the reduction of bandwidth demand as one of the goals of its Defense Information Infrastructure Common Operating Environment (DII COE). The DII COE is a set of standards pertinent to the Defense Department's hardware, software, and networking capabilities. A number of commercial companies are beginning to respond with more-capable software tools that control computer file server software on the basis of the available network bandwidth.

APPENDIX A

The Army's Current Communications Initiatives

The Army is currently pursuing three major programs that are expected to increase its communications bandwidth: the Warfighter Information Network-Tactical (WIN-T), the Joint Tactical Radio System (JTRS), and a new satellite communications (SATCOM) terminal. Those programs had formerly been considered part of the Army's digitization initiative, which—as discussed in Chapter 1—has now been subsumed under the Army's transformation effort.[1] Although no longer separately identified in the Department of Defense's annual reports to the Congress, the digitization initiative still comprises about 100 Army programs whose collective goal is to field advanced information technologies throughout the service's combat and support forces.

In the defense program covering fiscal years 2003 to 2007, overall spending for the service's digitization programs was projected to average $4.1 billion dollars annually.[2] Approximately 28 percent of that funding was allocated to communications; 24 percent to command-and-control software programs; 14 percent to digital processing capabilities in equipment such as the Abrams tank, Bradley fighting vehicle, and Comanche helicopter; 27 percent to data-link improvements in intelligence, reconnaissance, surveillance, and targeting systems; and 7 percent to training and development of doctrine for the new digital equipment. The WIN-T, JTRS, and new SATCOM terminal account for about one-third of the 28 percent of total digitization funding allocated to communications between 2003 and 2007. The remaining two-thirds of that funding is designated for upgrades to the Army's large fixed-base satellite communications terminals; for command-and-control programs associated with the fire-support and intelligence nets; and for other, less expensive, less complex radio systems.

Under the Army's current plans, the likely total investment in the WIN-T, JTRS, and new SATCOM terminal will range from $19 billion to $24 billion through 2020. Procurement of the satellite terminals will be completed by 2007, and the first combat units will be equipped with all three systems by 2010. From 2003 to 2009, the Army plans to spend roughly $2.5 billion on the JTRS, $1.7 billion on the WIN-T, and $200 million on the new SATCOM terminal.

The WIN-T Program

The overall purpose of the WIN-T program is to develop and purchase communications equipment for tactical operations centers (TOCs) at the brigade and higher levels of command. The program will provide equipment to replace the current mobile subscriber equipment phone system; it will also provide the radios (particularly such equipment as the high-capacity line-of-sight, or HCLOS, radio), the telephone switching networks and computers,

---

1. Some details related to the digitization initiative are reported in annual defense reports. See, for instance, Secretary of Defense William S. Cohen's *Annual Report to the President and Congress* in 1999, 2000, and 2001. Over that period, the Army's annual investment in digitization increased from $3 billion to $3.6 billion.

2. Congressional Budget Office, *The Long-Term Implications of Current Defense Plans* (January 2003).

the local area networks, and the control software that make up the communications element of a TOC.

WIN-T equipment will be capable of operating with the JTRS to allow the higher command levels to communicate with brigade and lower levels. Although the requirements for the WIN-T are still being developed, the operational data rate that its radios are expected to provide will be about 833 kilobits per second (Kbps)—the effective throughput at the applications layer for the HCLOS radio.[3] Such a data rate will allow streaming video at a speed of about 16 frames per second.

Because the requirements for the WIN-T are continuing to evolve, its costs are uncertain. The Army's current estimates of the program's total costs range from $4 billion to $9 billion through 2020; estimates by contractors range from $5 billion to $6.6 billion. Recent discussions within the Army's program office for the WIN-T have focused on issues related to bandwidth supply, and program officials have noted shortfalls that are consistent with the results of this Congressional Budget Office (CBO) analysis. One solution that the Army is considering is redundancy—that is, increasing from five to 101 the number of WIN-T sites (either on vehicles or in the TOCs) that military units of the future will have.[4] Still to be settled, however, are questions regarding the costs of that redundancy and the "footprint" (the amount of space devoted to computer and communications equipment in the TOCs) associated with such an expansion.

The Joint Tactical Radio System

The JTRS is a family of radios designed to provide interoperable tactical communications among the military services. Communications among the services' different radios during the 1991 Gulf War, and the JTRS program was initiated following that conflict. The military's procurement of the JTRS is divided into "clusters," which are differentiated according to the lead agency responsible for developing the radio hardware for various mission areas or weapon platforms (for example, a tank or a ship). (*Table A-1* summarizes the services' participation in the JTRS program by cluster, lead agent, and funding.)

The Army is most heavily involved in the Cluster 1 portion of the JTRS program. The cluster's goals are to develop and acquire a family of new software-programmable radios that can be used to communicate with a number of existing Army and Air Force radios operating in frequency bands of between 2 megahertz and 2 gigahertz. The radios built under Cluster 1 must also be capable of using a new wide-band networking waveform (WNW) that will provide substantially increased bandwidth compared with the amount provided today. The WNW is a family of four waveforms at different frequencies that offer different levels and types of capability.[5] The latest description of the Army's requirements for the WNW states:

> "[T]he WNW shall support greater than 2 Mbits per second of Network Throughput as a threshold. The WNW shall have the ability to make efficient use of extra frequency spectrum when available and shall support Network Throughputs of greater than 5 Mbits per second as an objective. . . . The JTRS WNW operating in the Point-to-Point mode shall support a user throughput rate of greater than 2 Mbps in each direction."[6]

On the basis of that description, CBO used 2 Mbps as the throughput rate for the JTRS Cluster 1 radio system in its analysis of bandwidth supply in 2010.

---

3. The maximum engineering throughput is about 8 million bits per second (Mbps), point to point.

4. Reported to the Congressional Budget Office in a briefing by the WIN-T Program Office titled "Warfighter Information Network-Tactical," February 20, 2003.

5. The WNW is being developed in stages: stage 1, a wide-band waveform available by 2004; stage 2, a midband waveform that has a low-probability-of-intercept (LPI) and a low-probability-of-detection (LPD) capability by 2005; stage 3, a midband waveform with "anti-jam" capability by 2005; and stage 4, a narrow-band, special-access waveform by 2006. The WNW's interoperability with the legacy waveforms is expected by 2006.

6. Department of the Army, JTRS Joint Program Office, *JTRS WNW Functional Description Document* (August 23, 2001), pp. 12-13.

Table A-1.
Planned Investment in the Joint Tactical Radio System, by Organization and Cluster

Cluster and Investors (Lead agency)	Mission Area or Platform	Planned Spending, 2003 to 2007 (Millions of dollars)
Cluster 1 (Army)	Vehicular/Rotary Wing	
Army		516
Marine Corps		51
Air Force		22
Cluster 2 (SOCOM)	Handheld Radios/Manpack	
SOCOM		10
Air Force		35
Cluster 3 (Navy)	Ships/Fixed Sites	517
Aviation Cluster (Air Force)	Fixed-Wing Planes	
Air Force		938
Navy		0
Subtotal		2,089
Joint Program Office	Waveform Development	269
Total		2,358

Source: Congressional Budget Office based on the Department of Defense's Future Years Defense Program for fiscal years 2003 through 2007.

Note: SOCOM = Special Operations Command.

The most recent estimates of costs for the JTRS are on a par with an average production cost per radio of about $127,000.[7] In comparison, the radios being replaced by the JTRS range in cost from about $8,000 to $28,000. However, per unit of bandwidth, systems like the SINCGARS (Single-Channel Ground and Airborne Radio System) and EPLRS (Enhanced Position Location Reporting System) cost about 25 cents to 50 cents per bit per second; the corresponding cost for the JTRS is projected to be about 6 cents, which is only slightly higher than that using fiber-optic cable.[8]

The Army plans to buy a total of 106,000 JTRS radios, which would be sufficient to equip about one-half of its forces. (Left unequipped would be most of the 15 enhanced separate brigades, their combat support and combat service support units, and the National Guard divisions.)[9] Under current plans, the JTRS program is projected to cost about $15 billion. Of that amount, procurement costs would account for $13.5 billion, and research and development costs, $1.5 billion. The Army currently plans to purchase about 10,000 JTRS radios per year, on average, at an annual cost of about $1 billion over the period from 2010 to 2020.

Those plans may be jeopardized, however, if perceptions of the program as a high-risk effort prove to be correct. (A recent analysis sponsored by the Army noted the high level of risk associated with the JTRS program's successful

---

7. Based on the December 2002 Selected Acquisition Report.

8. The *Report of the Defense Science Board Task Force on Tactical Battlefield Communications* (February 2000, p. 102) states that the cost of bandwidth using fiber-optic cable is 4 cents per bit per second.

9. The SINCGARS SIP (special improvement program), which procured the most recent and capable SINCGARS radio, purchased 108,000 of them. However, the number of fielded SINCGARS is about 211,000, which includes all versions of the radios that have been purchased.

completion.)[10] Elements that contribute to that assessment include the complexity of the software development required, the size and weight constraints imposed on the radios, the amount of power that they will consume, the heat that they dissipate, and interference problems that are anticipated among the waveforms when the radios are colocated. For example, each JTRS radio will have multiple central processing units and power amplifiers to cover the broad range of frequencies (2 megahertz to 2 gigahertz) over which each radio must operate.[11] That equipment will dissipate substantial heat into the confined spaces (such as tanks) in which the radios will be installed, making the reliability of their electronic components a primary requirement. The Army has identified potential solutions to such problems but has yet to demonstrate that they will work collectively.

Another risk factor associated with the JTRS program is the challenge of ensuring that the radios will function properly as part of a complex communications network. The JTRS will provide the communications capability to support a mobile battlefield version of what today is largely an Internet based at fixed sites.[12] The Defense Science Board has stated that "the Internetworking aspects of the program, a critical contribution of moving the DoD [Department of Defense] point-to-point and broadcast wireless infrastructure into an integrated Internetwork, is [sic] not being adequately addressed."[13]

Related cautions about the same set of issues were expressed as part of the supporting assessments generated for the Defense Acquisition Board's decisions in 2003 related to the Future Combat System (FCS).[14] The assessments noted the following: "Bandwidth could be an Achilles Heel. . . . Estimated requirements could reach 10's of Mbps for UA [this CBO study conservatively estimates 3 Mbps to 10 Mbps]. . . . [The] Army must have a credible bandwidth requirement and planned solution to pass Milestone B."[15] Those bandwidth-related concerns were reiterated by the recent Institute for Defense Analyses (IDA) assessment of the FCS.[16] The report noted that what it termed "critical enablers"—that is, complementary systems such as the JTRS and WIN-T—must be "managed and fielded on the same time schedule as FCS." While in general, the IDA study panel took the Army's technology assessments at face value, they did assign a yellow advisory assessment (implying caution) to the JTRS's "wideband waveforms."[17]

How would the results of CBO's analysis be affected if the JTRS program did not deliver its improved radios by 2010? The most likely substitutes for the JTRS would be either the NTDR (Near-Term Data Radio); a digital, improved EPLRS; or a new radio with a similar maximum bandwidth. If a substitution was necessary, the operational bandwidth provided by such a radio would be about 30 Kbps instead of the 200 Kbps provided by the JTRS. As a consequence, the mismatch in 2010 between bandwidth supply and demand in the Army's future

---

10. Department of the Army, Army Materiel Systems Analysis Activity [AMSAA], *Army Future Combat Systems Unit of Action Systems Book, Version 1.3(s)* (September 18, 2002), pp. 2-2, 2-18, and 2-19.

11. In principle, one processor could be sufficient. But the constraints of timely development plus the Army's desire to incorporate existing, patented technologies and use existing control systems force the choice of multiple processors.

12. In Operation Iraqi Freedom, for instance, the client/server Internet could be characterized as movable or mobile clients networking through static banks of fileservers located at fixed sites in locations such as Kuwait, Qatar, and later Baghdad.

13. Office of the Under Secretary of Defense for Acquisition, Technology, and Logistics, *Report of the Defense Science Board Task Force on Tactical Battlefield Communications* (February 2000), p. 93 and Annex D.

---

14. These materials are the 31 technology assessments referred to by E.C. Aldridge Jr. in his memorandum of May 17, 2003, to Secretaries of the military departments regarding the Future Combat Systems acquisition decision.

15. L. Delaney, "Independent Review of Technology Maturity Assessment for Future Combat Systems Increment 1" (March 3, 2003). The study was commissioned by the Army prior to the May 17, 2003, Defense Acquisition Board Milestone 2 meeting on the FCS.

16. L.D. Welsh, "Report of the Indendent Assessment Panel for the Future Combat System" (draft, Institute for Defense Analyses, April, 28, 2003).

17. Ibid., p. 60.

Table A-2.
Effective Bandwidth Supply Versus Peak Demand in 2010, with a JTRS Substitute

Command Level[a]	Relative Supply Versus Peak Demand (S : D)[b]
Corps	1 : 10 to 30
Division	1 : 10 to 30
Brigade[c]	1 : 3 to 10↑
	1 : 50 to 150↓
Battalion	1 : 15 to 30
Company	1 : 2 to 10
Platoon	1 : 0.5 to 1
Squad/Vehicle	1 to 2 : 1

Source: Congressional Budget Office.

Note: JTRS = Joint Tactical Radio System.

a. At the higher command levels, the table refers to the operations networks only. At lower levels, the distinctions between the various communications networks (for example, operations, intelligence, and fire-support) become less clear.

b. Based on an approximate logarithmic scale, the color coding is as follows: yellow indicates that supply is between about one-third and three times demand (a marginal demand/supply match), and orange signifies that demand is from three to 10 times supply. Red (here used for the corps, division, lower brigade-level, and battalion level relationships) means that demand exceeds supply by a factor of 10 or more.

c. The up-arrow (↑) indicates the throughput rate for communications to equivalent or higher command levels. The down-arrow (↓) indicates the throughput rate to lower command levels.

Objective Force would be much greater and more pervasive than the results discussed in Chapter 2 (*see Table A-2*).

The SATCOM Terminal

The Army is developing a new satellite communications terminal to replace the canceled STAR-T (Super High Frequency Triband Advanced Range Extension Terminal) program. It has several candidates, including a modified version of an existing SATCOM terminal being used by the Air Force, called the Lightweight Multiband Satellite Terminal (LMST), which operates in the C-, X-, and Ku-band frequencies. Those frequencies, which in the order given provide increasingly more bandwidth and increasingly lower probabilities of interception and detection, collectively enable communications with most military and commercial satellites. The Army expects that an improved version of the LMST, available around 2006, will provide about 2.5 Mbps of operational throughput. By 2010, with the addition of Ka-band capability, the terminal's effective throughput should be about 8 Mbps per channel.[18] The Army plans to buy 60 of these new satellite communications terminals at a cost of $195 million between 2003 and 2007.

---

18. Reported in *Inside the Army*, June 3, 2002, p. 1, and confirmed by discussions with the Army staff on August 8, 2002. There are Ku-band satellite terminals that generate such throughputs, but the Ka frequency band offers the additional advantage of a low probability of interception and detection.

APPENDIX B

Extrapolating Continuous-Flow Information Across Command Levels

To justify its assumption that demand for communications bandwidth increases by a factor of three between one level of command and the next higher level, the Congressional Budget Office (CBO) used data from a study conducted for the Army's Communications and Electronics Command (CECOM) in 2000 by the Mitre Corporation.[1] Table B-1 displays Mitre's estimates of bandwidth demand at the division, brigade, and battalion command levels; the estimates exclude voice-only communications that are not digital and therefore not relevant to CBO's analysis. The relationships between the commands and the structure of the communications networks associated with those estimates are assumed to exist through 2010 except as noted in Chapter 2. In addition to communications traffic associated with the operations network, the estimates in Table B-1 include bandwidth demand for intelligence data, fire-support data, and network management. Because CBO's analysis deals only with bandwidth demand for the operations net, however, those additional demands must be removed to facilitate comparisons between the two sets of estimates (see Table B-2).

The data in Table B-2 are consistent with CBO's assumption that the demand for communications bandwidth increases by a factor of three from one level of command to the next higher level and decreases by a factor of three at the next lower level. The data also indicate, however, that this assumption is only approximately true—factors somewhat higher or lower than three would also be consistent with the available data. Consequently, CBO has considered the effects on the results presented in Chapters 1 and 2 of using either a factor of two or a factor of four to extrapolate bandwidth demands to command levels above and below that of the division (see Tables B-3 and B-4 for a selection of these variations).

As the tables indicate, for extrapolations that use factors ranging from two to four, significant shortfalls in the supply of bandwidth are projected to occur at the brigade and battalion levels today and at the corps and division levels in 2010. The variations that are not presented (a factor-of-four extrapolation of demand in 2003 and a factor-of-two extrapolation for 2010) do not change those qualitative results.

---

1. Yosry Barsoum, "Bandwidth Analysis (ACUS Only) of Division Main, Maneuver Brigade TOC, and Tank Battalion" (briefing prepared for the Army's Communications and Electronic Command by the Mitre Corporation, February 29, 2000).

Table B-1.
Peak Bandwidth Demand at the Digitized Division, Brigade, and Battalion Levels in 2003, by Mission

(In bits per second)

Mission	Division TOC	Brigade TOC	Battalion TOC
Voice Communications	1,424,000	240,000	61,440
Maneuver	1,651,041	315,204	271,394
Fire Support	313,386	311,181	310,134
Intelligence and Electronic Warfare	2,570,327	1,364,168	283,763
Combat Service Support	297,118	128,158	0
Air Defense	30	197	100
Network Management	1,168,335	505,846	54,125
Total	**7,424,237**	**2,864,754**	**980,956**

Source: Congressional Budget Office based on briefing materials developed by the Mitre Corporation for the Army's Communications and Electronics Command, February 29, 2000.

Notes: The bandwidth demand detailed above is for the Amy Common User (communications system) and represents most of the operations net's trunk demand. However, it excludes some older radios and walkie-talkies that, for the most part, support only local, nondigital voice communications that are not relevant to CBO's analysis.

TOC = tactical operations center.

Table B-2.
Peak Bandwidth Demand for the Operations Nets at the Digitized Division, Brigade, and Battalion Levels in 2003, by Source of Demand

(In kilobits per second)

Source of Throughput Demand	Division	Brigade	Battalion
Telephone	1,400	300	100
Army Battle Command System			
Classified	300 to 1,000	300	30 to 100
Unclassified	100 to 300	100	0
Unmanned Aerial Vehicles	100 to 300	100 to 300	100 to 300
Video Teleconferencing	1,000	100 to 300	100 to 300
Total	**2,900 to 4,000**	**900 to 1,300**	**330 to 800**

Source: Congressional Budget Office.

Table B-3.
Effective Bandwidth Supply Versus Peak Demand in 2003 Using a Factor-of-Two Extrapolation, by Command Level

(In kilobits per second)

Command Level[a]	Relative Supply Versus Peak Demand (S : D)[b,c]
Corps	1 : 1.5 to 3
Division	1 : 5 to 8
Brigade[d]	1 : 1.5 to 3 ↑
	1 : 20 to 30 ↓
Battalion	1 : 10 to 20
Company	1 : 4 to 6
Platoon	1 : 2 to 3
Squad/Vehicle	1 : 8 to 14

Source: Congressional Budget Office.

a. At the higher command levels, the table refers to the operations networks only. At lower levels, the distinctions between the various communications networks (for example, operations, intelligence, and fire-support) become less clear.

b. Ranges have been extrapolated from the division level for the corps level and for the company level and below. They have been rounded down to the nearest power of three, which allows the demand per command level to be distinguished while maintaining a conservative (lower) estimate of the aggregate demand.

c. Based on an approximate logarithmic scale, the color coding is as follows: yellow indicates that supply is between about one-third and three times demand (a marginal demand/supply match), and orange signifies that demand is approximately three to 10 times supply. Red (used here for the lower brigade-level relationship and at the battalion level) means that demand exceeds supply by a factor of 10 or more.

d. The up-arrow (↑) indicates the throughput rate for communications to equivalent or higher command levels. The down-arrow (↓) indicates the throughput rate to lower command levels.

Table B-4.
Effective Bandwidth Supply Versus Peak Demand in 2010 Using a Factor-of-Four Extrapolation, by Command Level

Command Level[a]	Relative Supply Versus Demand (S : D)[b]
Corps	1 : 12 to 40
Division	1 : 10 to 30
Brigade[c]	1 : 2 to 6↑
	1 : 4 to 10↓
Battalion	1 : 1 to 4
Company	3 to 4 : 1
Platoon	5 to 8 : 1
Squad/Vehicle	10 to 20 : 1

Source: Congressional Budget Office.

Note: Ranges have been extrapolated from the division level for the corps level and for the company level and below. They have been rounded down to the nearest power of three, which allows the demand per command level to be distinguished while maintaining a conservative (lower) estimate of the aggregate demand.

a. At the higher command levels, the table refers to the operations networks only. At lower levels, the distinctions between the various communications networks (for example, operations, intelligence, and fire-support) become less clear.

b. Based on an approximate logarithmic scale, the color coding is as follows: green (used here for the company, platoon, and squad/vehicle relationships) means that supply is greater than demand by approximately a factor of three or more; yellow indicates that supply is between about one-third and three times demand (a marginal demand/supply match); light orange signifies that demand is approximately three times supply; orange indicates that demand is approximately three to 10 times supply; and red (used here for the corps and division relationships) means that demand exceeds supply by a factor of 10 or more.

c. The up-arrow (↑) indicates the throughput rate for communications to equivalent or higher command levels. The down-arrow (↓) indicates the throughput rate to lower command levels.

APPENDIX C

Compressing Data to Reduce Bandwidth Demand

Data compression has been suggested as a means to reduce bandwidth demand. This Congressional Budget Office (CBO) analysis concludes, however, that the most likely effect of pending advances in data compression will not be to empty saturated trunk communications lines. Instead, those lines will probably remain saturated with increasingly compressed data.

Data compression techniques differ depending on whether or not some losses of data can be tolerated. If they can, so-called *lossy* techniques may be used; if no losses may occur, *lossless* techniques must be employed. The transmission of picture images (so-called *imagery*) or streams of pictures (known as *streaming video*) can have some data errors, on the order of a few percent, because the human eye and brain unconsciously correct such anomalies. But compression techniques that are used for transmitting military orders, network management and other control information, situation assessments, and much of the rest of military data must be lossless.

A number of fast techniques exist for lossless compression, but the prospects are small for major improvements by 2010 in the amount of compression such techniques can achieve. The best one can obtain is about a 2:1 compression, on average, and military computer systems today routinely employ such compression techniques for transferring large data files. Computer users typically attain lossless compression (and decompression) when they use computer applications that zip, gzip, and unzip files. Improvements on those techniques are being pursued, but they usually involve the sequential application of known techniques. Those techniques may produce improvements of about 15 percent relative to current performance.

In contrast, substantial improvements in lossy compression techniques will occur by 2010 because of an expected reduction, by an additional order of magnitude, in the transmitted data throughput. A key to that transition will be the change from MPEG-2 to MPEG-4 standards for data compression. (MPEG stands for Moving Picture Experts Group.) Associated with each MPEG standard is a different numerical algorithm. The MPEG-2 uses numerical algorithms based on the fast Fourier transform, which was optimized for computers about 40 years ago to recast data associated with a stream of pixels—or more properly, the pixel indices. (The data are recast as a properly weighted sum of trigonometric sines and cosines and their higher harmonics, which are related to the second, third, and higher powers of sines and cosines.) The trigonometric functions associated with the recasting are called the basis functions. The fast Fourier transform exactly evaluates multiplicative weights for each basis function so that the weighted sum of the basis functions exactly equals the original data when evaluated at the pixel indices. In that sense, the weights replace the original data on a one-for-one basis.

However, the mathematics of Fourier transforms proves that smaller weights can often be ignored (set to zero) while the remaining weights allow the original data stream to be approximated to within acceptable error tolerances. The ratio of the retained weights relative to the original number of weights is the compression ratio. For typical two-dimensional imagery and video, acceptable error levels can be maintained by setting about nine-tenths of the weights to zero. Therefore, under the MPEG-2 standard, the retained weights for typical two-dimensional images imply a data compression ratio of about 10:1.

The new algorithm underlying the MPEG-4 is called the wavelet transform. In addition to working with basis functions like the sines and cosines noted above, it employs additional, specially tailored basis functions that allow minimal errors with even fewer retained weights. (In more metaphorical language, each basis function is called a wavelet because graphs of some of them look like waves.) Fast algorithms for employing the wavelet transform have been developed over the past 20 years. With the MPEG-4 standard, typical data compression ratios for two-dimensional images can often be expected to improve to about 100:1.

In the military context, the Air Force expects to begin using MPEG-4 data compression on a limited experimental basis in 2003, a move that would achieve an order-of-magnitude reduction in the bandwidth required for transmitting video images collected by unmanned aerial vehicles (UAVs). But rather than using the techniques to reduce the total demand for bandwidth, the Air Force plans to increase the resolution of the images transmitted or the number of sensors (or both), which would result in no net decrease in total bandwidth demand.[1] After the effectiveness of UAVs in the recent Iraq and Afghanistan campaigns, together with their novelty and the rapidly changing doctrine for their employment by the Army, there is no reason to expect that the Army will use these data compression techniques differently.

Therefore, although improvements in data compression will occur, CBO believes that in the future, the improvement will be used to keep the communications pipes full of more and increasingly compressed data rather than allowing the pipes to be emptied. Thus, improved data compression is unlikely to affect the results of CBO's analysis regarding the 2010 mismatch between bandwidth supply and demand on the battlefield.

---

1. Personal communication to the Congressional Budget Office from Col. Rhys MacBeth, Commander, Digital Imagery Video Compression and Object Tracking Battlelab, Eglin Air Force Base, Florida, August 10, 2002.

Glossary of Abbreviations

AAR	after-action report
ABCS	Army Battle Command System)
AFATDS	Army Forward Artillery Tactical Data System
AMSAA	Army Materiel Systems Analysis Activity
ASCII	American Standard Code for Information Interchange
ATCCS	Army Tactical Command and Control System
ATM	asynchronous transfer mode
AWE	advanced warfighting experiment
BLOS	beyond line of sight
C2	command and control
C4	command, control, communications, and computers
CECOM	Communications and Electronics Command
CSSCS	Combat Service Support Control System
DAWE	Division XXI Advanced Warfighting Experiment
DSL	digital subscriber line
DSCS	Defense Satellite Communications System
EPLRS	Enhanced Position Location Reporting System
EPLRS (VHSIC)	Enhanced Position Location Reporting System (Very High Speed Integrated Circuit)
FBCB2	Force XXI Battle Command, Brigade and Below
FRAGO	fragmentary order
FCS	Future Combat Systems
Gbps	gigabits per second
GCCS	Global Command and Control System
HCLOS	high-capacity line of sight
http	hypertext transfer protocol

ID	infantry division
IDA	Institute for Defense Analyses
IP	Internet protocol
IT	information technology
JTRS	Joint Tactical Radio System
Kbps	kilobits per second
LAN	local area network
LMST	Lightweight Multiband Satellite Terminal
LOS	line of sight
LPD	low probability of detection
LPI	low probability of intercept
Mbps	megabits per second
MIST	Multiband Integrated Satellite Terminal
MPEG	Moving Picture Experts Group
MSE	mobile subscriber exchange
NTC	National Training Center
NTDR	Near-Term Data Radio
OPORD	operational order
PCM	pulse-coded modulation
SATCOM	satellite communications
SDR	software-defined radio
SINCGARS	Single-Channel Ground and Airborne Radio System
SINCGARS (SIP)	Single-Channel Ground and Airborne Radio System (System Improvement Program)
SHF	super high frequency
SMART-T	Secure Mobile Anti-Jam Reliable Tactical-Terminal
STAR-T	Super High Frequency Triband Advanced Range Extension Terminal
SUAV	small unmanned aerial vehicle
TCP	transmission control protocol
TOC	tactical operations center
TUAV	tactical unmanned aerial vehicle
UAV	unmanned aerial vehicle
UTO	unit task order
VTC	video teleconferencing

GLOSSARY OF ABBREVIATIONS

WAN	wide-area network
WIN-T	Warfighter Information Network-Tactical
WNW	wide-band network waveform

www.ingramcontent.com/pod-product-compliance
Lightning Source LLC
Chambersburg PA
CBHW081902170526
45167CB00007B/3123